Laws and Lawmakers

MW00577198

Laws and Lawmakers

Science, Metaphysics, and the Laws of Nature

Marc Lange

OXFORD
UNIVERSITY PRESS

2009

OXFORD
UNIVERSITY PRESS

Oxford University Press, Inc., publishes works that further
Oxford University's objective of excellence
in research, scholarship, and education.

Oxford   New York
Auckland   Cape Town   Dar es Salaam   Hong Kong   Karachi
Kuala Lumpur   Madrid   Melbourne   Mexico City   Nairobi
New Delhi   Shanghai   Taipei   Toronto

With offices in
Argentina   Austria   Brazil   Chile   Czech Republic   France   Greece
Guatemala   Hungary   Italy   Japan   Poland   Portugal   Singapore
South Korea   Switzerland   Thailand   Turkey   Ukraine   Vietnam

Published by Oxford University Press, Inc.
198 Madison Avenue, New York, New York 10016

www.oup.com

Oxford is a registered trademark of Oxford University Press.

Library of Congress Cataloging-in-Publication Data
Lange, Marc, 1963–
Laws and lawmakers : science, metaphysics, and the laws of nature / Marc Lange.
p. cm.
Includes bibliographical references.
ISBN 978-0-19-532813-4; 978-0-19-532814-1 (pbk.)
1. Philosophy of nature.   2. Physical laws.   3. Philosophy and science.   I. Title.
BD581.L255 2009
113—dc22   2008041648

9 8 7 6 5 4 3 2 1
Printed in the United States of America
on acid-free paper

For Dina

The beauty of electricity, or of any other force, is not that the power is mysterious and unexpected, touching every sense at unawares in turn, but that it is under *law* . . .

— Michael Faraday (1858: 560)

We need scarcely add that the contemplation in natural science of a wider domain than the actual leads to a far better understanding of the actual.

— Arthur Eddington (1928: 266–67)

Preface

This book aims to answer two kinds of perennial philosophical questions about laws of nature:

1. *Questions about how laws differ from facts that do not qualify as laws,* such as the fact that the Andromeda galaxy is about 2.5 million light-years from Earth and the fact that each of the families living on my block has two children. Philosophers generally call these non-laws "accidents." These questions, then, concern the various respects in which laws differ from accidents—not merely how the *actual* laws and accidents differ, but how the laws would differ from the accidents no matter what particular laws and accidents there were. The topic is *lawhood:* the status of being a law rather than an accident. In what ways must the facts possessing lawhood differ from the facts lacking it?

2. *Questions about how the various differences between laws and accidents are to be explained.* Which of these differences are responsible for which others? Is there a *fundamental* difference between laws and accidents that ultimately accounts for all of the other respects in which they differ? What are the *lawmakers:* the facts in virtue of which the laws *are* laws rather than accidents?

Here, in summary, are the answers I shall give.

It has long been recognized that laws of nature differ from accidents by standing in a more intimate relation to "subjunctive facts": facts about what would happen under certain circumstances that may not actually come to pass. Appropriately, circumstances that do not arise are called "counterfactual" circumstances. A sentence concerning what would have happened under some counterfactual circumstance—such as "If I had gone shopping, then I would have bought a quart of milk"— is called a "counterfactual conditional" (or "counterfactual," for short). To begin to see how laws differ from accidents in their relation to counterfactuals, consider this example: Had Jones missed his bus to work this morning, then the actual laws of nature would all still have held, but some of the facts that are actually accidents (such as Jones's perfect on-time attendance record at work) would not still have held.

This approach to distinguishing laws from accidents, though illuminating, threatens to become disappointingly circular: the laws, unlike the accidents, would still have held under any counterfactual circumstance *that is logically consistent with the laws.* For instance, the laws are logically consistent with Jones's missing his bus to work this morning but not with Jones's accelerating to beyond the speed of light (since a law prohibits bodies from doing that). Accordingly, the laws would still have held even if Jones had missed his bus to work this morning, but not if Jones had accelerated to beyond the speed of light. Thus, to see how laws stand out from accidents by displaying greater persistence under counterfactual circumstances, we must begin by restricting our attention to those circumstances that are logically consistent with the laws. To make the laws stand out, we must first put the laws in!

This circularity may appear unavoidable. However, I will explain how to avoid it—and thus show how the subjunctive facts suffice to determine the laws. I will then bravely propose that the subjunctive facts *are* the lawmakers.

Instead of becoming entitled to offer this proposal by first critiquing every rival proposal that has been made, I will get right to the juicy part. In chapter 1, I will explain my solution to the circularity problem that I mentioned above. During the succeeding three chapters, I will occasionally look at the competing pictures of natural law offered by David Armstrong, Brian Ellis, and David Lewis—but mainly in the service of giving a crisp presentation of my own account and the arguments for it.

Here are some "coming attractions."

Chapter 1: Laws Form Counterfactually Stable Sets

In the first chapter, I argue that a few sets of truths possess a remarkable kind of invariance under counterfactual circumstances—an invariance that I call "stability." Roughly speaking, a set of truths is "stable" exactly when its members would all still have been true under any counterfactual circumstance that is logically consistent with their all being true. The set containing all and only the laws is a stable set. For example, the set of laws is logically consistent with Jones's missing his bus to work this morning, so the laws' stability requires that the laws would all still have been true if Jones had missed his bus to work this morning. In contrast, no set containing accidents is stable (unless, perhaps, it contains all truths). For example, take the set containing just the logical consequences of the truth that Jones always arrives at work on time. It is "unstable": it is logically consistent with the supposition that Jones missed his bus to work this morning, but its members would not all still have been true under that supposition.

The rest of this book springs entirely from this proposal (so it had better be mostly right!). It suggests a way to capture the laws' hierarchical structure: that certain laws transcend the idiosyncrasies of others in that they would still have held even if those others had been different. (For example, in classical physics, perhaps the laws of momentum, mass, and energy conservation are more general than the laws governing specific forces: energy would still have been conserved even if there had been different forces instead of gravity, electromagnetism, and so forth.) This proposal regarding the laws' characteristic relation to counterfactuals also has the welcome consequence that had Jones missed his bus to work this morning, the actual laws would not only still have been true, but also still have been laws.

Chapter 2: Natural Necessity

In chapter 2, I use ideas from chapter 1 to explain what makes laws necessary (sharply setting them apart from accidents), but not as necessary as "broadly logical" truths (such as conceptual, mathematical, metaphysical, and [narrowly] logical truths). Laws of nature have traditionally been thought to possess a distinctive species of necessity (dubbed "natural" necessity). For example, take the fact that any two positive (or negative) electric charges repel each other. Because this regularity holds as a matter of natural law, it is inevitable, unavoidable—necessary. An exception to it is (naturally) impossible. Any two like charges not only as a matter of fact *do* repel each other—they *must*. Yet the laws are also thought to be contingent truths; unlike the broadly logical truths, the laws of nature could have been different from the way they actually are.

The laws' apparent status as "contingent necessities" has often been considered paradoxical. Consequently, some philosophers ("scientific essentialists," such as Brian Ellis) have rejected the laws' contingency; they characterize laws as possessing the same strong variety of necessity as broadly logical truths do. Other philosophers ("Humeans," such as David Lewis), in contrast, have downplayed the laws' necessity, arguing that no profound metaphysical gap separates laws from accidents. In this chapter, I aim to do justice not only to the laws' necessity (by which they are distinguished from accidents), but also to their contingency (by which they are distinguished from broadly logical truths). I show how genuine varieties of necessity (such as logical necessity and natural necessity) are distinguished from mere "conversational" or "relative" necessities (as when Jones says, "I must be going now; I have to catch my bus to work"). By using the concept of "stability," I propose unpacking every genuine variety of necessity in terms of subjunctive facts. The varieties of necessity (logical, natural, and so forth) can then be understood as distinct species of the same genus. My proposal explains not only what makes natural necessity weaker than other genuine varieties of necessity, but also why all such necessities must stand in a definite ranking by strength in the first place.

Chapter 3: Three Payoffs of My Account

In chapter 3, I display three additional fruits of the account given in the previous chapters.

1. I explain what it would be for the laws to change (that is, for there to be temporary laws, as distinct from eternal but time-dependent laws). I argue that my account nicely explains why natural laws are immutable.

2. I consider how symmetry principles in physics may constitute "meta-laws"—that is, laws governing the laws governing what happens. I argue that the concept of "stability" can be used to elaborate the notion of a "meta-law" so that the meta-laws' relation to the laws they govern (the "first-order laws")

mirrors the first-order laws' relation to the facts they govern. This approach accounts for the meta-laws' modal status and the explanatory power commonly attributed to them, such as the capacity of symmetry principles to explain why various conservation laws hold.

3. I look at the special relation between laws and "objective chances"—as when an atom of the radioactive isotope polonium-210 is by law 50% likely to decay sometime during the next 138.39 days (the isotope's half-life). I argue that the laws' relation to objective chances falls naturally out of my account (whereas it must be inserted into Lewis's account by hand).

Chapter 4: A World of Subjunctives

In this final chapter, I aim to answer the question: Are the laws necessary by virtue of being laws, or are they laws by virtue of being necessary? It seems to me that their necessity is ultimately what makes them laws— what sets them apart from accidents. Since their necessity is constituted by subjunctive facts, I conclude that the lawmakers are subjunctive facts. However, subjunctive facts are widely regarded as very poor candidates for being ontologically prior to laws. Indeed, the way that counterfactual conditionals (which express subjunctive facts) manage to be true is notoriously murky. *My boldest suggestion in this book is that subjunctive facts are ontologically primitive and responsible for laws,* a view that is contrary to the traditional conception of laws as "underwriting" or "supporting" counterfactuals. I offer several additional arguments for my view.

1. A theory according to which essences, universals, or other heavy-duty metaphysics is responsible for both laws and subjunctive facts cannot account nicely for the laws' characteristic relation to counterfactuals. The laws' stability would have to be inserted into such an account in an ad hoc manner. The laws' relation to counterfactuals has a much more straightforward explanation if subjunctive facts are the lawmakers.

2. It has long been recognized that laws have a distinctive power to explain why various facts hold by rendering those facts (naturally) necessary. (For example, all like charges in fact mutually repel because

this regularity *must* hold.) Now take the facts (whatever they are) in virtue of which the fundamental laws are laws. If those lawmakers are not necessary, then they are unable to bestow necessity upon the laws, and so the laws lose their explanatory power. On the other hand, if the lawmakers are necessary, then what makes them so? If their necessity is constituted by other facts, then are those facts necessary or not? If they are necessary, then the regress continues, but if they are not necessary, then the laws' necessity is again compromised. On my picture, the various subjunctive facts that serve as lawmakers, unlike other subjunctive facts, are (naturally) necessary. Each of them has its necessity constituted by other subjunctive facts that also help to make the same laws. (Each of those other subjunctive facts is necessary by virtue of yet other subjunctive facts that help to make those same laws, and so forth.) That is why the laws are able to render certain regularities necessary without deriving their own necessity from anywhere else, much less from facts that are unnecessary.

3. Instantaneous rates of change (such as velocity, according to classical physics) figure in the universe's state at a given moment. Indeed, a quantity's instantaneous rate of change at time t traditionally plays various causal and explanatory roles. But its capacity to do so cannot be accounted for by the standard reductive analysis of this rate in terms of a mathematical function of the changing quantity's values at various times throughout t's neighborhood. The best way to account for the causal and explanatory roles played by some quantity's instantaneous rate of change at t is to interpret that rate in terms of an irreducibly *subjunctive* fact. (For example, in classical physics, for a body at t to have an instantaneous speed of V centimeters per second is for it to be the case that *were* the body (existing at t) to remain in existence after t, the body's trajectory *would* have a time-derivative from above at t equal to V cm/s.) Thus, the universe's state at a given moment cannot be purged of irreducibly subjunctive facts. If such facts must be countenanced anyway, parsimony urges us to put them to work as the lawmakers.

★   ★   ★

In an earlier book, *Natural Laws in Scientific Practice* (Lange 2000), I remained steadfastly neutral about whether laws help to make counterfactual conditionals true, or the reverse, or whether some third kind of fact is responsible for both laws and subjunctive facts. Now I am prepared to argue that subjunctive facts are the lawmakers—that this view best explains the laws' relation to counterfactuals. However, one could reject this proposal and still accept a considerable portion of what I have to say.

Although the central idea in chapter 1 was embedded (though not especially highlighted) in my earlier book, most of the present book is new—or, at least, lifted from articles I wrote after that book appeared:

"Must the Fundamental Laws of Physics Be Complete?" *Philosophy and Phenomenological Research* 78 (2009): 312–45.

"Why Do the Laws Explain Why?" in *Dispositions and Causes,* ed. Toby Handfield (286–321). Oxford: Oxford University Press, 2009.

"Why Contingent Facts Cannot Necessities Make," *Analysis* 68 (2008): 120–28.

"Could the Laws of Nature Change?" *Philosophy of Science* 75 (2008): 69–92.

"Laws and Meta-Laws of Nature: Conservation Laws and Symmetries," *Studies in History and Philosophy of Modern Physics* 38 (2007): 457–81.

"How to Account for the Relation between Chancy Facts and Deterministic Laws," *Mind* 115 (2006): 917–46.

"Do Chances Receive Equal Treatment under the Laws? Or: Must Chances Be Probabilities?" *British Journal for the Philosophy of Science* 57 (2006): 383–403.

"How Can Instantaneous Velocity Fulfill Its Causal Role?" *Philosophical Review* 114 (2005): 433–68.

"Reply to Ellis and to Handfield on Essentialism, Laws, and Counterfactuals," *Australasian Journal of Philosophy* 83 (2005): 581–88.

"A Note on Scientific Essentialism, Laws of Nature, and Counterfactual Conditionals," *Australasian Journal of Philosophy* 82 (2004): 227–41.

Some important topics in my earlier book (such as lawhood's relation to inductive confirmation, laws as an elite class of natural necessities, natural kinds, and "ceteris-paribus" laws of inexact sciences) make no appearance here.

Many thanks to my good friends John Roberts and John Carroll, who have supplied me with a wealth of valuable feedback during recent years. I am immensely grateful for their generous help and their friendship. Thanks also to Jamin Asay, Gabriele Contessa, Adam Cureton, Crawford Elder, Katie Elliott, Mathias Frisch, Alan Hajék, Chris Haufe, Chris Pearson, and Matthew Slater, who read some or all of the manuscript and helped me to improve it. My title is an affectionate nod to David Armstrong's *Truth and Truthmakers* (2004).

The philosophy department at the University of North Carolina at Chapel Hill has given me an incredibly congenial and stimulating environment in which to work and to play. For that, I am especially grateful to the department chair (and my dear friend), Geoffrey Sayre-McCord. I also wish to record my gratitude to the anonymous referees for the journals in which the above papers appeared. The care that referees typically take to provide authors with worthwhile feedback makes me proud to belong to the philosophical community.

My greatest debts are to my wonderful family: my wife, Dina, and my children, Rebecca and Abe. For their love, patience, joy, and encouragement, I am grateful beyond words.

Contents

Laws and Lawmakers

Laws Form Counterfactually Stable Sets

1.1. Welcome

I am sitting in the waiting room of a car-repair shop, pounding out these words on my laptop computer while waiting for my car to be fixed. In a host of ways, the laws of nature impinge upon my situation. The auto mechanics are relying on their knowledge of various natural laws in trying to explain why my car is misbehaving and to predict whether it would stop doing so if some of its components were adjusted or replaced. My laptop's cooperative behavior arises from the electrons in its circuits obeying various laws that the laptop's designers were justified in predicting they would obey. Those electrical engineers designed the laptop's insides as they did, rather than according to some other blueprint, because they knew that it would not have worked had they designed it in that other way. Presumably, in using the laws of nature to predict how laptops of various possible designs would behave, the engineers took it for granted that the actual laws would still have been laws even if the laptop's innards had been different. The same applies to the auto mechanics who are relying upon their knowledge of the natural laws to tell them how various possible repairs would affect my car. By the same token, the electrical engineers and auto mechanics also presumably believe that the actual laws would still have been laws even if computers and cars had never been invented—or, for that matter, even if humanity had never evolved or the sun had never formed.

These examples illustrate the most important roles that laws of nature play in science: in connection with scientific explanations, predictions of the future, and "counterfactuals" (that is, "predictions" of what would have happened under certain circumstances that do not actually come about). Whereas scientists aim to discover what the laws of nature are, philosophers aim to identify what it is to *be* a law of nature—in other words, what *makes* a given fact qualify as a law rather than an "accident." For example, philosophers aim to specify what it is about reality in virtue of which it is a law that all gold objects are electrically conductive but an accident (if it is true at all) that all gold cubes are smaller than one cubic mile—to take Hans Reichenbach's famous example.[1] What it is to be a natural law must account for the work that natural laws do in connection with scientific explanations, predictions, and counterfactuals.

In this book, I propose an account of what natural laws are that explains why they do what they do. Admittedly, I lack any sort of knockdown argument for my proposal. Nevertheless, I think that my account has some novel and even (dare I say it?) elegant features that make it attractive.

Although I will not comprehensively survey the other accounts of natural law that are currently on the market, I will occasionally highlight some respect in which my proposal contrasts with its chief rivals. I will also present a few general recipes for generating worries about other proposed accounts of natural law. My own proposal is designed to avoid these worries. Ready? OK, then—let's begin.

1.2. Their Necessity Sets the Laws Apart

Laws of nature have traditionally been contrasted with two other kinds of facts: accidents and "broadly logical" truths. What separates these three kinds of facts?[2]

Let's start with the accidents. The word "accident" here is a bit of potentially misleading philosophical jargon. Please do not confuse it with the ordinary meaning of "accident"—what I mean when I say to you, "Our meeting here was no accident; I was looking for you," or when the owner of a car dealership confidentially informs us, "It is no accident that every car in my showroom smells so fresh; I put the same chemical in each of them, to give them all that fabled 'new car smell.'" To call some events "accidental" in this ordinary sense is roughly to say that they were unintentional ("I accidentally spilled my soup"), that they were just a coincidence ("I encountered him by accident; I wasn't trying to find him"), or that there is no common explanation from which they all spring. On the other hand, in philosophical discussions of natural law, an "accident" can be no coincidence. Indeed, an accident typically has an explanation. For example, that every car in the showroom has the same "new car" smell is no coincidence (the dealer has just told us the explanation), and it certainly was intentional on the dealer's part, but it is still an "accidental" truth in the philosophical sense—the sense relevant to this book. An "accident" in that sense is simply a truth that does not follow from the natural laws (and the "broadly logical" truths) alone. In other words, an accident could have failed to hold without any violation of the natural laws. For example, no natural law has to be violated for the showroom to contain a car without the "new car" smell.

Take another example: suppose that many apples are hanging on the tree in my backyard, and all of them are now ripe. Their ripeness is an accident, since even if some of them were not ripe, the laws of nature could still all have held. If the warm weather had arrived a few weeks later, for instance, then the apples would not yet have been ripe, though the natural laws would have been no different. Nevertheless, it is no coincidence that every single one of those apples is ripe today. Their ripeness resulted from the recent weather conditions, the levels at which various plant hormones have been flowing through the tree, and so forth. Since each of these factors was common to every one of those apples, they all ripened together. Certain laws of nature governing chemical reactions are also responsible for the apples' ripeness. These laws determined how the weather, the plant hormones, and so forth influenced the rate at which the apples ripened. Again, that there are laws and other conditions explaining why all of those apples are now ripe does not keep this fact from qualifying as an accident. Those other conditions are themselves accidental; there is no explanation of the ripeness of any of those apples that appeals to no accidents at all, but exclusively to laws of nature.

Let's now contrast the laws of nature with the "broadly logical" truths. A broadly logical truth possesses a kind of *necessity* that is possessed neither by natural laws nor by accidents. For instance, one kind of broadly logical truth consists of the mathematical truths, such as the fact that there is no way to divide 23 evenly by 3. There does not merely *happen* to be no integer that added to itself, and then added again, equals 23—in the way that there merely happens to be no gold cube larger than a cubic mile (and even in the way that like charges merely happen to repel rather than to attract). Rather, there *couldn't* have been an integer that added to itself, and then added again, equals 23. That it is *impossible* to divide 23 evenly by 3 explains why no one has ever succeeded in figuring out a way to do so, no matter how much mathematics she knows—and why every time someone tries to divide 23 objects evenly into thirds, she fails. None of these efforts could have succeeded. They all fail because they must; their failure was inevitable. Analogous considerations apply to other kinds of broadly logical truths, such as conceptual truths (for example, "All sisters are female"), narrowly logical truths ("Either all emeralds are green or some emerald is not green"), and metaphysical truths ("Red is a color" or perhaps "Water is H_2O").

Just as each broadly logical truth possesses some variety of necessity that accidents and natural laws lack, so likewise there is a species of necessity possessed by every natural law but by none of the accidents. An accident (even one that is not an utter coincidence, such as that all of the apples on my tree are ripe) just happens to obtain. A gold cube larger than a cubic mile could have formed, but (I presume) the requisite conditions happened never to arise. In contrast (again following Reichenbach), it is no accident that a large cube of uranium-235 never formed, since the laws governing critical masses and nuclear chain-reactions prohibit it. (Actually, they merely render it extremely unlikely, but let's ignore this detail for the sake of Reichenbach's nice example.) In short, things *must* conform to the laws of nature. The laws *prohibit* perpetual-motion machines. It is not simply that all material objects accelerated from rest *as a matter of fact* fail to exceed the speed of light. Rather, they *could not* have exceeded light speed; their failure to outpace light is inevitable, unavoidable—necessary. As a popular tee-shirt says, the speed limit of 3×10^8 meters per second isn't just a good idea; it's "THE LAW."

That a characteristic variety of necessity is possessed by the natural laws, setting them apart from ordinary, run-of-the-mill facts, has been recognized for as long as the concept of a natural law has been explicitly invoked. Here, for example, is Richard Hooker way back in 1593, explaining how obedience to the natural laws is compulsory: "[T]hings naturall...doe so necessarily observe their certaine Lawes, that as long as they keepe those formes which give them their being they cannot possibly be apt or inclinable to do otherwise then they do."[3] A thing's "forme," Hooker says, enrolls it in a natural "kinde" and explains its "working." According to Hooker, talk of "formes" can be translated into talk of "lawes": a thing's behavior is explained by the laws applicable to the kind of thing it is.

That the laws, by virtue of their necessity, possess an explanatory power that accidents lack was famously codified about 350 years later by Carl Hempel's "covering law" models of scientific explanation. Whatever the shortcomings of Hempel's overall account of explanation, its motivation in the differences between laws and accidents is easy to appreciate.[4] For example, a certain powder burns with yellow flames, rather than flames of any other color, because the powder is a sodium salt and it is a law that all sodium salts, when ignited, burn with yellow flames. (This law, in turn, is explained by more fundamental laws.) The powder *had* to burn with yellow flames considering that it was a sodium salt; this "had-to-ness" expresses the laws' distinctive kind of necessity. In contrast, we cannot explain why my wife and I have two children by citing the regularity that all of the families living on our block have two children—since this regularity is accidental. Were a childless family to try to move onto our block, it would not encounter an irresistible opposing force or automatically acquire two children upon moving in. Like a natural law, a mathematical truth possesses explanatory power by virtue of its necessity. For example, the fact that 23 cannot be divided evenly by 3 explains why it is that every time mother tries to divide 23 strawberries equally among her three children without cutting any (strawberries), she fails.[5]

Although laws possess a variety of necessity, it is widely believed that laws are *not* as necessary as the broadly logical truths. It has generally been thought that the natural laws could have been different (as they are in certain science-fictional universes, where starships travel beyond light speed), whereas the broadly logical truths are a great deal less malleable. (How *could* there have been a round square?) The status of the natural laws thus sounds paradoxical: they are contingent necessities, falling somewhere between the broadly logical truths and the accidents. Faced with the laws' apparently anomalous(!) character, some philosophers (such as Brian Ellis and other "scientific essentialists") have tried to argue that the laws are in fact every bit as necessary as the broadly logical truths. Other philosophers (such as David Lewis and other "Humeans") have taken the opposite approach, arguing that the laws are not separated from the accidents by a profound metaphysical gap. In contrast to both of these approaches, my account will recognize the laws' distinctive, intermediate status.

In what does the laws' necessity consist? To answer this question is one of the main goals of this book, since to do so would be to understand what laws of nature are. Laws are set apart from accidents by the necessity they possess and from broadly logical truths by the necessity they lack. An account of the laws' necessity should reveal not only how laws differ from accidents and from broadly logical truths, but also how the laws' "natural necessity" qualifies as a kind of *necessity*—as a species of the same genus as the variety (or varieties) of necessity possessed by [narrowly] logical, conceptual, mathematical, and metaphysical truths.

1.3. The Laws' Persistence under Counterfactuals

Whereas a large gold cube could have formed, it is inevitable, unavoidable—necessary—that all bodies accelerated from rest fail to exceed light's speed. Had Bill Gates wanted to build a large gold cube, then (I dare say) there might well have been a gold cube exceeding a cubic mile.[6] But even if Bill Gates had wanted to arrange for a body to be accelerated from rest to beyond the speed of light, no body would have been so accelerated. Even if Bill had possessed 23rd-century technology, he would have failed had he tried to accelerate a body from rest to beyond the speed of light.

In elaborating the laws' necessity, I have just made use of "counterfactual conditionals" (or "counterfactuals," for short): statements of the form "Had p been the case, then q would have been the case" (symbolically: $p \;\Box\!\!\rightarrow q$, where I will refer to p as the "counterfactual supposition" or "antecedent" and to q as the "consequent").[7] For example: Had we tried to build a perpetual-motion machine, the law of energy conservation would have prevented our success. We cannot get around the natural laws; they are unavoidable. Counterfactuals are used to express the fact that the laws would still have held even if various other things had been different. The laws of nature govern not only what actually happens, but also what would have happened under various circumstances that did not actually come to pass.

Counterfactual conditionals, though asserted by each of us every day, are notorious in philosophy. To begin with, they seem to violate rules of reasoning that look sensible and are obeyed by many other kinds of conditionals. For instance, $p \ \square\!\!\rightarrow q$ and $q \ \square\!\!\rightarrow r$ do not logically entail $p \ \square\!\!\rightarrow r$. As an illustration, let $p \ \square\!\!\rightarrow q$ be "Had Socrates been a woman, then Socrates would not have been a philosopher." That's true, considering the mores of ancient Greek society. (Of course, a woman in ancient Greece could have lived "a life of the mind," and some did. But evidently and regrettably, she could not have been a "professional" philosopher.) Now let $q \ \square\!\!\rightarrow r$ be "Had Socrates not been a philosopher, then Socrates would still have been a man." Of course, that's true, too. But $p \ \square\!\!\rightarrow r$ is "Had Socrates been a woman, then Socrates would still have been a man," which is obviously false.

Another notorious feature of counterfactuals is that it is mysterious what makes (some of) them true. It is quite clear which features of the world are responsible for making it true (if it is true) that all gold cubes are smaller than a cubic mile: namely, the sizes of the various gold cubes throughout the universe's history. But it is much harder to identify the features of the world in virtue of which it is true (if it is true) that had Julius Caesar been in command during the Korean War, then he would have used the atomic bomb.[8] Nevertheless, the standard view is that such a counterfactual is made true by certain features of the actual world, such as Caesar's hotheaded personality. But how, in general, are the responsible features of the world to be picked out? Nelson Goodman famously grappled with this problem.[9] He began with the attractive thought that "Had the match been struck, it would have lit" is made true by various natural laws along with the fact that the match is actually dry, oxygenated, well made, and so forth. However, no matter how many facts about the actual match Goodman included, he found that they did not suffice to make the counterfactual conditional obtain. To them must be added the truth of various other *counterfactual conditionals:* that had the match been struck, it would *still* have been dry, oxygenated, and so forth. It is difficult to see how the original counterfactual's truth can be reduced entirely to noncounterfactual facts (and the laws of nature), though there have been many ingenious attempts to solve this problem.[10]

Another feature of counterfactuals that is widely considered suspect is their context-sensitivity. The counterfactual "Had Caesar been in command during the Korean War, then he would have used the atomic bomb" is true in some contexts, whereas in other conversational environments, the counterfactual "Had Caesar been in command during the Korean War, he would have used catapults" is true instead. Which features of the actual world are preserved under a given counterfactual supposition (that Caesar was in command during the Korean War) and which are allowed to vary (Caesar's personality or his knowledge of armaments) depends to some extent upon our interests in entertaining that supposition. If we are in the midst of illustrating Caesar's gung-ho personality, then the point of the counterfactual is to contribute to that discussion, and so it is true in that context that had Caesar been in command during the Korean War, he would have used the atomic bomb. Of course, not every counterfactual supposition is relevant in every context. If we are trying to describe Caesar's personality, then to ask, "What weapons would Caesar have used in the Korean War, had Caesar been more cautious and less ambitious?" is utterly beside the point.

Although counterfactuals are context-sensitive, violate attractive-looking logical principles, and are not straightforwardly made true by features of the actual world, counterfactuals are not utterly disreputable. There are strict logical principles regulating their use, even if those principles are not quite the most familiar ones. For instance, $p \:\square\!\!\rightarrow q$ and $(p \& q) \:\square\!\!\rightarrow r$ logically entail $p \:\square\!\!\rightarrow r$. (The Socrates example where $p \:\square\!\!\rightarrow q$ and $q \:\square\!\!\rightarrow r$ are true, but $p \:\square\!\!\rightarrow r$ is false, does not violate this principle, since $(p \& q) \:\square\!\!\rightarrow r$ is "Had Socrates been a woman and a nonphilosopher, then Socrates would still have been a man," which is false.) Even young children have little trouble figuring out whether certain counterfactuals are true. Without giving it a second thought, we routinely assert such counterfactuals as "Had I not gotten lost along the way, I would have arrived here sooner." We are sensitive to the context-sensitivity of counterfactuals just as we easily grasp which sorts of remarks are relevant in a given conversation. Science ascertains that various counterfactuals are true, as when Lavoisier discovered that someone who is standing up and moving about would have consumed less oxygen had she instead been sitting quietly.

Indeed, our observations confirm the truth of various counterfactuals just as they confirm various predictions about the actual world.

Our past observations of emeralds confirm not only that all of the actual emeralds lying forever undiscovered in some far-off land are green, but also that had there been emeralds in my pocket right now, then my pocket would have contained something green. (It is not self-evident which of these predictions is more "remote" from our observations.) When we confirm that my pocket would contain something green were there emeralds in it, that confirmation is unaffected by whatever evidence we may have regarding whether there actually are any emeralds in my pocket. So in confirming that my pocket would contain something green were there emeralds in it, we may be confirming both a prediction about the actual world and a counterfactual conditional. Claims about what *would have been* are confirmed right along with claims about what actually *is*.

(A claim like "Were there emeralds in my pocket, then my pocket would contain something green" is a "subjunctive conditional" rather than a counterfactual, since it fails to presuppose that there are no emeralds in my pocket. The corresponding counterfactual conditional is "Had there been emeralds in my pocket, then my pocket would have contained something green." The connection between the antecedent and consequent of a subjunctive conditional that is true is presumably the same as the connection between the antecedent and consequent of a counterfactual conditional that is true. I shall use the symbol "$p \ \square \!\!\rightarrow q$" to represent both subjunctive and counterfactual conditionals, and I shall often use the term "counterfactuals" to encompass both.)

Although a given counterfactual conditional may have different truth-values in different contexts, this phenomenon is hardly confined to counterfactuals. Claims with indexicals ("I," "now") and demonstratives ("this") display the same behavior. Plausibly even some claims without indexicals or demonstratives do, too: how close Jones's height must be to exactly six feet, in order for "Jones is six feet tall" to be true, differs in different contexts. One possible explanation of these phenomena is that the same sentence expresses different propositions on different occasions of use. Claims with indexicals and demonstratives certainly appear to do so without provoking undue suspicion.

In chapter 4, I shall say more about the facts expressed by counterfactuals. Fortunately, we do not need to have a philosophical account of the truth-conditions of counterfactuals in order to be entitled to use counterfactuals as we ordinarily do: to have (in a given context) great confidence in the truth of certain counterfactuals and the falsehood of certain others. Our goal in this chapter is to identify precisely how laws differ from accidents in their relation to counterfactuals. Having done so, we will then be in a good position to understand the laws' characteristic species of necessity. (That will be our aim in chapter 2.)

1.4. Nomic Preservation

Many examples suggest that laws and accidents stand in different relations to counterfactuals.[11] Had Jones missed his bus to work this morning, for instance, then every actual law of nature would still have held, but some of the actual accidents (such as that Jones always arrived at work on time) would not still have held. This example suggests that an accident's range of invariance under counterfactual suppositions is *narrower* than a law's—in other words, that for any law and any accident, the range of counterfactual suppositions under which the law is preserved wholly contains and goes somewhat beyond the range of counterfactual suppositions under which the accident is preserved. Accidents are thus more "delicate" than laws—more easily broken.

However, this thought is incorrect. Suppose a large collection of electrical wires, all of which are made of copper, are lying on a table. For the sake of the wires' utility, it is a good thing that copper is electrically conductive. Had copper been electrically insulating, then the wires on the table would not have been much good for conducting electricity. Now look at what just happened: under the counterfactual supposition that copper is an insulator, the law that all copper objects are electrically conductive obviously fails to be preserved. But the accident that all of the wires on the table are *made* of copper *is* preserved (at least in certain, easily imagined conversational contexts). To repeat: Had copper been electrically insulating, then the wires on the table, being made of copper, would have been useless for conducting electricity. So it is false that an accident's range of invariance under counterfactual suppositions is strictly narrower than a law's. There are counterfactual suppositions under which (in a given conversational context) a given accident is preserved but a given law is not.[12]

This example illustrates another important fact: although accidents may in some respect be more delicate than laws in having less resistance to being overthrown by counterfactual insults, a mere accident may nevertheless possess plenty of resilience. It is no law that all of the families living on our block have two children, yet this accident would still have held even if I had failed to brush my teeth this morning or worn an orange shirt today. Here is another accident possessing considerable persistence under counterfactual suppositions: whenever the gas pedal of my car is depressed by x inches and the car is on a dry, flat road, then the car's acceleration is given by the function $a(x)$. Let's call this accident "g" for "gas pedal." (Of course, g is no coincidence; whenever the gas pedal is depressed, the same facts about the car's internal makeup help to explain its acceleration. But since those facts are accidents, g is accidental, too.) My knowledge that g would still have held, had the gas pedal on a certain occasion been pressed down a bit farther, has been relevant many times recently to the guidance I have given my daughter, Rebecca, in teaching her how to drive my car.[13]

Let's find a better way to capture the difference between laws and accidents in their persistence under counterfactual suppositions. The copper-wire example required a counterfactual supposition under which the law that all copper objects are electrically conductive fails to be preserved. I resorted to the brute-force approach: "Had copper been electrically insulating." It would have sufficed to use a counterfactual supposition about the electron-band structure of copper atoms or about the behavior of electrons or about the operation of electric fields. Nevertheless, each of these suppositions would have to be like my original brute-force supposition in being logically inconsistent with *some* natural law (even if logically consistent with the law that all copper objects are electrically conductive).[14] A counterfactual supposition must contradict some law or other in order for it to undercut the law that every copper object is electrically conductive. In contrast, for any accident, we can find a counterfactual supposition that is logically *consistent* with all of the laws, but under which that accident fails to be preserved. For instance, no law is violated by Bill Gates wanting to have a large gold cube built, yet under this supposition, the accidental generalization about gold cubes might not still have held. This suggests the following idea, which I call "Nomic Preservation" (NP):

NP m is a law if and only if m would still have held under any counterfactual (or subjunctive) supposition p that is logically consistent with all of the laws (taken together).

In other words, m is a law exactly when $p \;\square\!\!\rightarrow m$ is true for any p that is logically consistent with the laws (taken all together). Nomic Preservation allows an accident to be invariant under a wide range of counterfactual suppositions—even under a supposition that contradicts laws. Yet NP still manages to distinguish laws from accidents.

However, NP requires several important refinements. They will occupy our attention for the rest of this section.

Let's start with an easy one. In a given conversational context, only certain counterfactual antecedents are relevant, considering our interests there. For example, suppose that several emergency room physicians are discussing whether the victim of a motor vehicle accident, who has just died, might have survived under various counterfactual circumstances: had she suffered only certain injuries without others, or had she been wearing a seat belt, or had the ambulance arrived at the accident scene sooner. In that context, presumably, counterfactual antecedents such as "Had human evolutionary history been different so that our vital organs were arranged differently" or "Had the human aorta been constructed out of steel" are irrelevant. The physicians in that context are concerned with human anatomy as it actually is, not as it might have been.

A counterfactual conditional with an antecedent that is irrelevant in a given context is perhaps neither true nor false in that context. Therefore, even if m is a law and p is logically consistent with all of the laws (taken together), it may be that $p \;\square\!\!\rightarrow m$ is not true in a given context because p is irrelevant there.[15] NP will have to be refined to leave room for this possibility.

Furthermore, even if m is an accident, it may be that *in a given context,* $p \;\square\!\!\rightarrow m$ holds for all counterfactual antecedents p that are relevant in that context and logically consistent with the laws. For example, suppose that I have just driven from Chapel Hill to Myrtle Beach in order to meet you, but I have arrived 30 minutes late. We discuss whether I would (or at least might) have arrived on time had I departed Chapel Hill an hour earlier, or had I taken U.S. Highway 15 instead of Interstate Highway 95, or had there been no accident to slow traffic on I-95, and so forth. You might ask whether I would have arrived any earlier had I turned left at a given intersection, and we might conclude that had I turned left, then I would have arrived even later because I would then have entered the ramp onto I-95 north (away from Myrtle

Beach) rather than south (toward Myrtle Beach). In this familiar sort of conversation, a counterfactual antecedent such as "Had Myrtle Beach been 100 miles nearer to Chapel Hill" is irrelevant. In this context, under every *relevant* counterfactual antecedent, the locations of Chapel Hill and Myrtle Beach and the routes taken by various highways are preserved. Our concern in this context is how I might have arrived on time, given the actual distances to be traveled and the roads actually available. Nevertheless, the locations of Chapel Hill and Myrtle Beach and the routes taken by various roads are accidents, not laws.

Their accidental character is reflected in the fact that there are *other* contexts where these facts are *not* preserved under counterfactual antecedents that are relevant there and logically consistent with the laws. For example, there are contexts where "Had the fall line been 150 miles farther inland in South Carolina, then I-95 would have been constructed farther from the coast there" is true, not to mention contexts where "Had the earth been only 40 million miles from the sun, then human beings would never have evolved, and so I-95 would not have been constructed" is true.

Although NP should allow an accident to behave like a law in *certain* contexts, NP should require that for any accident, there is *some* context where it does *not* behave like a law. Thus:

> *m* is a law if and only if for any conversational context, and for any *p* that is relevant as a counterfactual antecedent in that context and logically consistent with all of the laws (taken together), the proposition expressed in that context by "*p* $\square\!\!\rightarrow$ *m*" is true.[16]

Let's now look at another way in which NP must be refined. Even if NP succeeds in distinguishing laws from accidents, NP fails to distinguish laws from broadly logical truths. If the laws are preserved under every counterfactual supposition in a given range, then the broadly logical truths (which have an even stronger variety of necessity than the laws) may well be preserved there, too.

However, although the broadly logical truths are not *merely* natural laws, they are *at least* natural laws. They possess whatever variety of necessity the laws possess and more. So they are "by courtesy" counted among the laws. After all, if it is a law that burning hydrogen in oxygen yields only water, and it is a law that burning hydrogen in oxygen yields only H_2O, then (if every logical consequence of laws is a law) it is a law that water is H_2O—although (some philosophers say) this is a metaphysical truth, reflecting water's *essence*.[17] Likewise, if it is a law that the speed of light is 3×10^8 meters per second and a law that the speed of light is half of 6×10^8 meters per second, then (if every logical consequence of laws is a law) it must be a law that $3 \times 10^8 = \frac{1}{2} \times 6 \times 10^8$, although this is a mathematical truth. NP can succeed in distinguishing laws from accidents only if the broadly logical truths qualify as laws.

There are motives for denying that every logical consequence of laws is a law. For instance, although it is a law that all emeralds are green and a law that all rubies are red, is it really a law that all things that are emeralds or rubies are green or red? (Presumably the reason why the stone in the King's ring is green or red is not because it is a ruby or an emerald, but rather because it is a ruby; that is why it is red, and hence green or red.)[18] A logical consequence of laws that is not itself a law will be preserved under every counterfactual supposition under which the laws entailing it are preserved. So no principle like NP can distinguish the laws proper from any of their logical consequences that are not laws. All of the laws' logical consequences hold "as a matter of law" even if not all of them are laws, narrowly speaking.

For the sake of simplicity, let's stipulate that as our default for the rest of this book, we shall interpret "natural law" expansively so that it includes all of the broadly logical truths as well as all of the logical consequences of laws. Over the course of this chapter and the next, we will see how the broadly logical truths differ from the other laws of nature by their greater invariance under counterfactual suppositions. On the other hand, I will have nothing to say here about how laws like "All emeralds are green" and "All rubies are red" differ from laws like "All things that are emeralds or rubies are green or red."

Now let's turn to the counterfactual antecedents under which NP demands that the laws be preserved: every counterfactual supposition p that is logically consistent with all of the laws. The antecedent must be logically consistent with all of the laws *taken together*, not merely with each law individually. In other words, there must be a logically possible world where p and all of the laws hold.

But for the laws to *hold* there, is it enough that every law (that is, every n where it is a law that n) is *true* there, or must they also be *laws* there? Which of these interpretations we place on NP can make a big difference. Suppose p is that *it is not a law* that energy is conserved. Obviously p is not logically consistent with the conjunction of all truths of the form "It is a law that n," since one of these truths is that it is a law that energy is conserved. But p is logically consistent with the bare fact that energy is conserved (in other words, with the fact that the total quantity of energy is the same at every moment) together with the truth of every other law. In a possible world where every actual law is true but p is also true, the total quantity of energy remains the same at every moment *as a matter of accidental fact.*

However, although this is *a* possible world where p is true, this is not the *closest* possible world where p is true—by which I mean simply that this is not what would have happened, had p been true. Rather, energy would *not* still have been conserved, had p been true (that is, had the laws not made energy conservation mandatory). Had there been no law prohibiting perpetual-motion machines, then scientists might well have built one by now. By the same token, had the laws not imposed 3×10^8 meters per second as a cosmic speed limit, there presumably would have been bodies accelerated from rest to beyond that speed— perhaps by the Stanford Linear Accelerator cranked up to full power. So NP is false if it demands that every law would still have been true under any counterfactual supposition that is logically consistent with the *truth* (though perhaps not with the *lawhood*) of all of the laws.

One way to avoid this problem is to refine NP so that it covers only counterfactual suppositions p where p is "sub-nomic," thereby excluding "Had energy conservation not been *a law*."[19] To explain this approach, let's start with the facts that we are trying to partition into laws and accidents.[20] Put aside all of those facts that could themselves hold at least partly by virtue of which facts are laws and which are not. The survivors include the fact that all emeralds are green (a law) as well as the fact that all gold cubes are smaller than a cubic mile (an accident)—but not the facts that it is *a law* that all emeralds are green, that it is *not a law* that all gold cubes are smaller than a cubic mile, and that no *laws* privilege any particular moment or location. Call the survivors the "sub-nomic facts," and let the "sub-nomic claims" be the claims that in any possible world are true (or false) exclusively by virtue of the sub-nomic facts there—not by virtue of which facts are laws and which are not. In other words, a claim is "sub-nomic" exactly when in any possible world, what makes it true (or false) there does not include which facts there are laws there and which are not. Note that as I shall use the term "sub-nomic," a sub-nomic claim, such as "Like charges repel" (or "All gold cubes are smaller than a cubic mile"), can nevertheless qualify as a law (or an accident).[21] That like charges repel is sub-nomic and a law, though the *fact that it is a law* is not a sub-nomic fact.

Let me also stipulate that the definition of "sub-nomic" treat the other species of modality in the same way as it treats natural modality. For example, just as "It is a law that like charges repel" is not sub-nomic, so likewise "It is a (broadly) logical necessity that all triangles have three sides" is not, but "All triangles have three sides" is. Sub-nomic claims do not contain such modal terms as "actually" or implicit references to other "possible worlds."[22]

The sub-nomic facts are, roughly speaking, the facts that laws might "govern" but that cannot themselves concern the composition of the "government." Accordingly, I take the sub-nomic facts to include facts about single-case objective chances, since science treats these facts as governed by laws just like facts about, say, the distribution of electric charge. For example, it is a sub-nomic fact that every atom of polonium-210, at each moment it exists, has a 50% chance of surviving for the next 138.39 days (the isotope's half-life). But the fact that this is a law is not a sub-nomic fact. It is a sub-nomic fact that every atom now in a given vial (labeled "polonium-210") has a 50% chance of surviving for the next 138.39 days. But the fact that this is an accident is not a sub-nomic fact.[23]

The sub-nomic facts lie at the base of a hierarchy of facts, where the facts on a given rung of the hierarchy are "governed" by some of the facts one rung higher. A given rung of the hierarchy consists of the facts (at least partly) about what's governing the facts on the rung immediately below (see fig. 1.1). Let's start three rungs up in the hierarchy. The facts there include that it is a law that the laws one rung below (the "first-order laws") exhibit various symmetries: roughly speaking, that they privilege no locations or moments, that they are the same in every reference frame in a certain family (as demanded by Einstein's "principle of relativity"), and so forth. The facts on this rung, then, specify what laws govern the first-order laws. (Such "meta-laws" will be discussed in chapter 3.) One rung below, on the second rung of the hierarchy, are the facts that the meta-laws govern, such as the facts specifying the laws that exhibit those symmetries (for example, the fact that it is a law that all electrons have negative electric charge and the fact that it is not a law that all gold cubes are smaller than a cubic mile). These facts are concerned not with what governs other *laws,* but rather with what governs the facts on the rung just below, which is the lowest rung of the hierarchy. The facts on the bottom rung are not about what governs some other facts, since there is nothing lower in the hierarchy to govern. These facts are governed but govern nothing themselves. They are the sub-nomic facts. The laws governing the sub-nomic facts are insufficient to determine all of the sub-nomic facts, since they do not entail the "initial conditions" (and even the laws together with the initial conditions fail to determine all of the sub-nomic facts, if the laws are statistical). Likewise the meta-laws are insufficient to determine the first-order laws; they merely impose some constraints on them.

That all like charges repel is a sub-nomic fact, whereas that it is a law that all like charges repel is not sub-nomic; it belongs on the second rung. Likewise, that no locations or moments are privileged by first-order laws belongs on the second rung, whereas that it is a law

Figure 1.1   The sub-nomic facts (tiny figure at extreme right) being bossed around by the laws governing them (second from right), which in turn are being constrained by the meta-laws (middle). I am not prepared to say for how many levels this hierarchy actually extends; that is for science to discover. The sub-nomic facts have no facts to boss around.

that first-order laws are so symmetric belongs on the third rung. That all first-order laws possess a certain feature may describe the facts governing the sub-nomic facts without itself explaining any sub-nomic fact (unlike the fact that it is a law that all like charges repel, which explains why in fact all like charges repel).

Let's now use the notion of "sub-nomic claims" to refine NP. If we take NP as concerned only with counterfactual suppositions p where p is sub-nomic, then we exclude from consideration the counterfactual suppositions that were causing trouble for NP, such as "Had energy conservation not been *a law.*"

To reduce clutter, let's henceforth reserve lower-case italicized English letters (such as p and m) for sub-nomic claims and leave mostly tacit the various other qualifications that I have just introduced. So we have arrived at

NP m is a law if and only if in any context, $p \,\square\!\!\rightarrow m$ holds for any p that is logically consistent with all of the n's (taken together) where it is a law that n (that is to say, for any p that is logically consistent with the first-order laws).

I endorse this principle.

1.5. Beyond Nomic Preservation

Although NP tells us something important about how laws differ from accidents in their relation to counterfactuals, we will see in the next two sections that we can go much further: by generalizing NP, we will see how the laws can be separated from the accidents *solely* by which counterfactuals are true. Before turning to that challenge, however, I want to identify a few further ideas that are suggested by the same sorts of thoughts that motivated NP. These ideas will prove fruitful later in this chapter. But I shall not add them officially to NP. It will be more convenient to reserve "NP" for the principle that I have just given.

The idea that some truth m is "preserved" under a given counterfactual supposition p was supposed to be captured by the fact that $p \,\square\!\!\rightarrow m$—in other words, by the fact that m would still have been true, had p been the case. But what if not only m, but also $\sim m$ (that is, m's negation) would have held, had p been the case? That's not really what we had in mind by m's "preservation"! Of course, not only would energy still have been conserved, had I tried to build a perpetual-motion machine (that is: $p \:\square\!\!\rightarrow m$), but also it is *not* the case that had I tried to build a perpetual-motion machine, then energy would *not* have been conserved (that is: $\sim (p \:\square\!\!\rightarrow \sim m)$). However, there may be more exotic counterfactual suppositions under which (at least in some conversational contexts) both m and $\sim m$ obtain, such as "Had there been a round square" or "Had some object been entirely made of rubber and also entirely made of copper." For $\sim m$ to hold along with m under some counterfactual supposition would be a disappointingly cheap way for m to be "preserved" there. Here is a means of capturing the idea that under various counterfactual suppositions p, the laws are preserved, but *not* in this cheap way:

> m is a law if and only if in any context, $p \:\square\!\!\rightarrow m$ and $\sim (p \:\square\!\!\rightarrow \sim m)$ hold for any p that is logically consistent with all of the n's (taken together) where it is a law that n.

We can shorten this principle a bit by replacing the *would*-conditionals in it with a *might*-conditional. I have already introduced some might-conditionals. For instance, I said that had Bill Gates wanted to build a large gold cube (p), then there might have been a gold cube exceeding a cubic mile (m)—symbolically: $p \:\Diamond\!\!\rightarrow m$. There are two important connections between might-conditionals and would-conditionals. Firstly, if it is not the case that $\sim m$ might have held, had p held, then m would have held, had p held. In other words, $\sim (p \:\Diamond\!\!\rightarrow \sim m)$ logically entails $(p \:\square\!\!\rightarrow m)$. Secondly, if it is not the case that $\sim m$ might have held, had p held, then it is not the case that $\sim m$ would have held, had p held. In other words, $\sim (p \:\Diamond\!\!\rightarrow \sim m)$ logically entails $\sim (p \:\square\!\!\rightarrow \sim m)$.[24] By virtue of these two might-would connections, both $p \:\square\!\!\rightarrow m$ and $\sim (p \:\square\!\!\rightarrow \sim m)$ in the indented principle above follow from $\sim (p \:\Diamond\!\!\rightarrow \sim m)$. Perhaps, then, we should consider the principle

> m is a law if and only if in any context, $\sim (p \:\Diamond\!\!\rightarrow \sim m)$ holds for any p that is logically consistent with all of the n's (taken together) where it is a law that n.

This looks like a good way to capture the laws' preservation.

Let me set this principle aside temporarily and turn to another idea suggested by the same sort of examples that motivated NP. As

I mentioned earlier, had the laws not required that energy be conserved, then energy might not have been conserved. This seems closely related to the thought that the reason why energy is in fact conserved is because its conservation is required by law; energy is conserved because it is a law that energy is conserved. (Here we have a covering-law explanation: That m is a law explains why m is the case by making m inevitable, unavoidable—necessary.) Now according to NP, energy would still have been conserved had p been the case, as long as p is logically consistent with the first-order laws. In the "closest p-world," then, why is it the case that energy is conserved? What is the scientific explanation there for the fact that energy is conserved? Presumably, the closest p-world is like the actual world in that energy is conserved there not by accident, but because a law requires it: energy conservation is a law in that world, too. If it is not the case that had someone tried to build a perpetual-motion machine, energy conservation would still have been a *law*, then why is it the case that had someone tried to build a perpetual-motion machine, energy would still have been conserved? Without the principle of energy conservation remaining a *law* under this counterfactual supposition, there is no need for the principle to remain *true* under that supposition.

That the laws of nature would have been no different, had Jones missed his bus to work this morning or Bill Gates wanted a large gold cube built, seems intuitively plausible and can be captured by this principle:

> m is a law if and only if in any context, "Had p been the case, then m would have been a law" holds for any p that is logically consistent with all of the n's (taken together) where it is a law that n.

Let's now explore one step further. If it is actually a law that m, then according to the above principle, $p \;\square\!\!\rightarrow (m$ is a law), and by NP, (m is a law) entails that $q \;\square\!\!\rightarrow m$—as long as p is logically consistent with the first-order laws, and q is likewise. Hence, if m is a law, then $p \;\square\!\!\rightarrow (q \;\square\!\!\rightarrow m)$ for any such p and q, and likewise no matter how many layers of counterfactuals are nested (or "embedded").

Please do not confuse the nested counterfactual $p \;\square\!\!\rightarrow (q \;\square\!\!\rightarrow m)$ with $(p \& q) \;\square\!\!\rightarrow m$. Their difference is especially evident when p and q are broadly logically inconsistent. For example, consider the nested counterfactual "Had the object been entirely made of rubber, then here's a counterfactual conditional that would have been true: had it been entirely made of copper, it would have been electrically conductive." That is true but plainly not equivalent to "Had the object been entirely made of rubber and been entirely made of copper, then it would have been electrically conductive." The nested counterfactual $p \; \Box \!\!\rightarrow (q \; \Box \!\!\rightarrow m)$ is not equivalent to $(p\&q) \; \Box \!\!\rightarrow m$ even when p is broadly logically consistent with q. For example, suppose that you and I run a race, I win, and I would always win were I to try. Had you won, then had I tried, I would have won. This nested counterfactual is obviously not equivalent to the false conditional "Had you won and I tried, then I would have won."

That the laws would still have been true, even under nested counterfactual antecedents, can be captured by the principle:

> m is a law if and only if in any context, $p \; \Box \!\!\rightarrow m$ holds, $p \; \Box \!\!\rightarrow$ $(q \; \Box \!\!\rightarrow m)$ holds, and so forth, as long as p is logically consistent with all of the n's (taken together) where it is a law that n, q is likewise, and so forth.

For instance, we believe that had Jones missed his bus to work this morning, then the natural laws would still have been exactly the actual laws and so (by NP) if Jones, after missing his bus, had done nothing about getting to work but click his heels and make a wish to materialize instantly at his office, he would not have gotten to work. That was a nested counterfactual that just went by. It concerned whether a given counterfactual conditional ("Had Jones done nothing but click his heels and make a wish to materialize instantly at his office, he would have gotten to work") would have been true under a certain counterfactual supposition ("Had Jones missed his bus to work this morning"). This nested counterfactual is covered by our latest principle.

Likewise, had Bill Gates tried to accelerate a body beyond the speed of light, then he would have failed, and moreover (here comes the nested counterfactual) had he access to 23rd-century technology, then had he tried to accelerate a body beyond the speed of light, he would still have failed. (Even if he had access to 23rd-century technology, it would still have been a law that no body is accelerated to beyond light speed.) On the other hand, it is merely an accident that all hurricanes rotate counterclockwise in the Northern Hemisphere. This regularity's accidental character is reflected in the truth of this nested counterfactual: had the

Earth rotated westward instead of eastward, then had there now existed some hurricanes in the Northern Hemisphere, they would all have been rotating clockwise.[25]

A bit earlier, I suggested that to require that the laws be preserved, but not in a cheap way, we can demand not only that p □→ m, but also that ~ (p □→ ~m), where both of these counterfactuals follow from ~(p ◊→ ~m). The same argument applies to preservation principles that include nested counterfactuals. If the preservation principle requires that q □→ (p □→ m), then in order to rule out the possibility that (p □→ m)'s preservation under q is of the cheap kind, the principle should also require that it *not* be the case that (p □→ m) might have been lost had q obtained. In other words, it should require that ~ (q ◊→ ~ (p □→ m)) hold. Now one of the connections we saw between might-conditionals and would-conditionals was that ~ (p ◊→ ~m) logically entails (p □→ m). In other words, ~ (p □→ m) logically entails (p ◊→ ~m). Therefore, if it is not the case that (p ◊→ ~m) might have held, then it is not the case that ~ (p □→ m) might have held. Hence, our new requirement's ~ (q ◊→ ~ (p □→ m))— in other words, that it is not the case that ~ (p □→ m) might have held, had q held—follows from ~ (q ◊→ (p ◊→ ~m)).

This last expression has other convenient implications as well. By another application of the same might-would connection, ~ (q ◊→ (p ◊→ ~m)) entails (q □→ ~ (p ◊→ ~m)), which, by one final application of the same might-would connection, entails (q □→ (p □→ m)). That was the first nested counterfactual that we incorporated into a preservation principle.

What we have just seen, then, is that ~ (q ◊→ (p ◊→ ~m)) entails every other result involving nested counterfactuals that we wanted to include in a preservation principle. Expressions like ~ (q ◊→ (p ◊→ ~m)) are all we need to use in order to construct a preservation principle that is powerful enough to encompass all of the various conditionals we want:

m is a law if and only if in any context,
  ~ (p ◊→ ~m),
  ~ (q ◊→ (p ◊→ ~m)),
  ~ (r ◊→ (q ◊→ (p ◊→ ~m))),...

all hold, as long as p is logically consistent with all of the n's (taken together) where it is a law that n, q is likewise, r is likewise, and so forth.

To keep things simpler, I have refrained from building these additional details officially into NP. But the principle at which we have just arrived succeeds in cashing out some further ideas that are suggested by the same thoughts that motivated NP. These ideas will prove valuable shortly.

1.6. A Host of Related Problems: Triviality, Circularity, Arbitrariness

Let's return to NP:

> NP m is a law if and only if in any context, $p \ \Box \!\!\rightarrow m$ holds for any p that is logically consistent with all of the n's (taken together) where it is a law that n.

NP and the other principles I have just mentioned accord well with our routine scientific practice of using the natural laws to ascertain what would have happened under various hypothetical circumstances. Intuitively, the laws supply a control panel of knobs for setting the universe's initial conditions (or any system's boundary conditions), and these knobs can be twisted (hypothetically!) in any fashion whatsoever that is logically consistent with the laws. No matter the setting to which the knobs are turned (counterfactually) within these generous limits, the actual laws would still have held.[26]

One entertaining example of knob-twisting takes place in Arthur Upgren's *Many Skies: Alternative Histories of the Sun, Moon, Planets, and Stars* (2005). An astronomy professor at Wesleyan and Yale, Upgren explains that in thinking about what the world would have been like under various counterfactual circumstances (such as had the solar system been located in a star cluster), "I have...not changed the laws of physics." Upgren takes what we believe the laws to be and extrapolates from them to the conditions that would have prevailed under various counterfactual circumstances.[27]

Principles like NP (though without some of the qualifications and elaborations that I have introduced) have been advanced by a host of philosophers.[28] They have also been contested by some philosophers. I relegate further discussion of these objections to an endnote[29] because we have an even bigger problem to worry about. Despite all of our work refining NP, we must face the fact that even if NP is true, it cannot reveal how the laws are set apart from the accidents by their relation to counterfactual conditionals.

One source of concern is that in one direction, NP is trivial. It is obvious that no accident would still have been true under every p that is logically consistent with the first-order laws, since if m is an accident, one such p is $\sim m$, and $\sim m \ \Box\!\!\rightarrow m$ is plainly false (at least when $\sim m$ is broadly logically possible). The trouble with NP's truth being trivial in this direction is that we might have expected "All gold cubes are smaller than one cubic mile" to reveal its accidental character not by failing to be preserved under "Had there been a gold cube larger than a cubic mile" (its failure to be preserved under *that* antecedent is surely not its fault!) but rather by failing to be preserved under a counterfactual supposition with which it is logically consistent, such as "Had Bill Gates wanted a large gold cube to be constructed."

NP is trivial in one direction because the range of counterfactual suppositions falling under NP encompasses exactly the p's that are logically consistent with the first-order laws. So for each accident m, the range includes $\sim m$, whereas for each law m, the range does not include $\sim m$. This bias toward the laws takes us to the heart of the problem with NP: it gives preferential treatment to the laws, allowing them to determine which counterfactual suppositions get to be considered. This amounts to the laws stacking the deck in favor of themselves.

Let me explain NP's "circularity" a little more carefully. NP uses the laws to pick out the range of counterfactual suppositions that, in turn, it uses to pick out the laws. This means that NP cannot distinguish the laws from the accidents solely on the basis of the truth-values (in all conversational contexts) of all of the counterfactual conditionals $p \ \Box\!\!\rightarrow m$. Rather, for NP together with the truth-values of those counterfactuals to reveal which sub-nomic facts are laws, the laws must first be distinguished on some *other* grounds, thereby picking out the range of counterfactual suppositions p that are logically consistent with the laws. Only then can NP pick out the laws among the sub-nomic facts as exactly the sub-nomic facts that are preserved under all of those suppositions.

But that's not all. Even if NP is true, NP cannot explain why the laws' distinctive sort of persistence under counterfactual suppositions makes the laws *special* or *important;* it cannot reveal the sense in which the laws bear an especially intimate relation to counterfactuals. Once again, the source of the trouble is that the concept of natural law appears in NP on both sides of the "if and only if"; NP allows the laws themselves to delimit the range of counterfactual perturbations under which a fact must be invariant in order for it to qualify as a law. Therefore, NP can portray the laws as special, in virtue of their invariance under this range of counterfactual suppositions, only if this particular range of counterfactual suppositions is itself special already. But this range is distinguished only by its ranging over exactly those suppositions that are logically consistent with all of the laws. Hence, NP can reveal what makes the laws special, as far as their invariance under counterfactual suppositions is concerned, only if there is some independent reason why the laws are special. Unless there is already some reason why this particular range of counterfactual suppositions is special, the laws' invariance under this range fails to reveal anything special about the laws.

NP's circularity is closely related to NP's triviality as far as accidents are concerned. NP permits the gold cubes accident to be invariant under a wide range of counterfactual suppositions. NP insists only that there be some counterfactual supposition p, logically consistent with every law, under which the gold cubes regularity would not still have held. Plainly, there is: Had there been a gold cube larger than a cubic mile! NP regards the failure of the gold cubes regularity to be preserved under this p as showing that the regularity is not a law, but only because this p is logically consistent with the laws—that is, only because the gold cubes regularity is not a law! There's the circle again.

It seems arbitrary to privilege the counterfactual suppositions that are logically consistent with the laws. We could, it seems, just as well have privileged the counterfactual suppositions that are logically consistent with, say, the fact that George Washington was the first president of the United States. But the facts that would still have held, under every one of *those* counterfactual suppositions, should not merit our attention in the way that the laws of nature should. (They have no special explanatory power, for instance.) What is so noteworthy, then, about preservation under one range of counterfactual suppositions as opposed to some other range? Once again, although NP may be true, it leaves unexplained why the laws are especially significant. For NP to tell us why the laws are special, we would have to know already what is special about being invariant under every counterfactual supposition that is logically consistent with (wait for it!) the laws, and so we would have to know already why the laws are special.

A few philosophers have mentioned a problem along these lines (though without necessarily elaborating it in terms of triviality, circularity, or arbitrariness) and despaired of ever resolving it.[30] I shall resolve it now. I shall spend the rest of this book trying to squeeze every last drop of philosophical juice out of this single move. So it had better be good!

1.7. Sub-nomic Stability

NP's problems arose from its giving special treatment to a range of counterfactual suppositions designed expressly to suit the laws. What if the same courtesy that NP gives the laws were extended to every set of sub-nomic truths? NP says that the laws would still have held under every counterfactual supposition that is logically consistent with the laws. So let's consider whether some set of sub-nomic truths containing accidents would still have held under every counterfactual supposition that is logically consistent with that set. We thus avoid arbitrarily privileging the range of counterfactual suppositions that is logically consistent with the laws. Rather than allowing the laws to dictate to any set of truths the range of counterfactual suppositions under which that set's invariance is to be assessed, let's allow each set to pick out for itself a range of counterfactual suppositions that it finds comfortable.

For example, take the set containing exactly the sub-nomic claims that are logical consequences of "All gold cubes are smaller than one cubic mile." Are this set's members all preserved under every sub-nomic counterfactual supposition that is logically consistent with all of them (taken together)? Of course, the set's members are *not* all preserved under the counterfactual supposition "Had there existed a gold cube larger than a cubic mile." But that supposition is not logically consistent with the set. To see that the set's members are not all preserved under every counterfactual supposition that is logically consistent with the set, notice that they are not all preserved under the supposition that Bill Gates wants a large gold cube to be constructed. Although that supposition *is* logically consistent with the set (and so its members *could* all still have held under it), they *wouldn't* all still have held under it.

This approach, I shall argue, allows us to distinguish the set of first-order laws (that is, the set containing exactly the sub-nomic truths m where it is a law that m) from any set of sub-nomic truths that contains some (but not all) of the accidents—while avoiding the problems of triviality, circularity, and arbitrariness afflicting NP. I will call the key property "sub-nomic stability." Roughly speaking, a set of sub-nomic truths is "sub-nomically stable" if and only if whatever the conversational context, the set's members would all still have held under every sub-nomic counterfactual (or subjunctive) supposition that is logically consistent with the set—even under however many such suppositions are nested. In other words, for any member m of the set, and for any sub-nomic suppositions $p, q, r, \ldots$, each of which is logically consistent with the set, all of the counterfactuals $p \ \square\!\!\rightarrow m, q \ \square\!\!\rightarrow (p \ \square\!\!\rightarrow m)$, $r \ \square\!\!\rightarrow (q \ \square\!\!\rightarrow (p \ \square\!\!\rightarrow m)) \ldots$ hold in every context. Moreover, as we saw in section 1.5, we should preclude *cheap* preservation of the set's members, which we can do by also requiring that it not be the case that their negations might have held under these suppositions—for instance, by requiring not only that $p \ \square\!\!\rightarrow m$, but also that $\sim (p \ \Diamond\!\!\rightarrow \sim m)$. We found that nested might-conditionals entail all of the conditionals we need. I will now put those nested might-conditionals to use in defining "sub-nomic stability":

> Consider a nonempty set Γ of sub-nomic truths containing every sub-nomic logical consequence of its members. Γ possesses *sub-nomic stability* if and only if for each member m of Γ (and in every conversational context),
>
> $\sim (p \ \Diamond\!\!\rightarrow \sim m),$
> $\sim (q \ \Diamond\!\!\rightarrow (p \ \Diamond\!\!\rightarrow \sim m)),$
> $\sim (r \ \Diamond\!\!\rightarrow (q \ \Diamond\!\!\rightarrow (p \ \Diamond\!\!\rightarrow \sim m))), \ldots$
>
> for any sub-nomic claims $p, q, r, \ldots$ where $\Gamma \cup \{p\}$ is logically consistent, $\Gamma \cup \{q\}$ is logically consistent, $\Gamma \cup \{r\}$ is logically consistent, ...

The motivations behind NP suggest that the set of first-order laws is sub-nomically stable. I shall call this set "Λ"—that is, "lambda" (for "law"). In contrast, the set containing exactly the sub-nomic logical consequences of the gold cubes generalization fails to be sub-nomically stable (since some members of the set might have been false, had Bill Gates wanted a large gold cube to be constructed).[31]

I shall argue that there is no sub-nomically stable set containing an accident—except perhaps the set of *all* sub-nomic truths. According to many proposed logics of counterfactuals, $p \;\Box\!\!\rightarrow\; q$ is true trivially whenever $p\&q$ is true (a principle known as "Centering"), and likewise for nested counterfactuals. If Centering is correct, then each member of the set of all sub-nomic truths is trivially preserved under every sub-nomic supposition p that is true. There are no sub-nomic suppositions p that are false and logically consistent with the set. (If p is a false sub-nomic supposition, then $\sim\!p$ is a member of the set.) In that case, the set of all sub-nomic truths trivially possesses sub-nomic stability. Accordingly, I will argue that Λ is the largest *nonmaximal* set that is sub-nomically stable. (On the other hand, if Centering is false, then even the set of all sub-nomic truths may lack sub-nomic stability. In fact, I believe that Centering fails in a universe where there are objective chances (other than 0 and 1), but perhaps it holds in a universe where there aren't.[32] But the truth of Centering need not concern us now; our focus is on the proposal that laws differ from accidents in belonging to a sub-nomically stable set that does not contain every sub-nomic truth.)

This proposal for distinguishing laws from accidents avoids the circularity afflicting the idea that the laws are the truths that would still have held under every counterfactual supposition that's logically consistent with the laws. Sub-nomic stability does not start by giving special privileges to the laws. It is very egalitarian; it does not grant the laws the right to dictate to every set the range of counterfactual suppositions under which that set's invariance is to be tested. Stability thus has the potential to be a genuinely special feature of the laws. Whether sub-nomic stability can realize this potential is the subject of the rest of this chapter (and indeed this book).

Let me allay one concern that you may have at this point. I have asked you to think about whether various counterfactuals are true. In trying to evaluate those counterfactuals, you may have found yourself thinking about what the laws of nature are. For instance, in considering whether there might have been a large gold cube had Bill Gates wanted one to be constructed, you may have said to yourself, "Well, I guess there might then have been a large gold cube, since it is just an accident that all gold cubes are smaller than a cubic mile." Likewise, in thinking about whether there might have been a perpetual-motion machine had Bill Gates wanted one to be constructed, you might have said to yourself, "No, there wouldn't have been, even if Bill had wanted one, since the laws of nature prohibit perpetual-motion machines." Accordingly, you may be worried that insofar as we are using our beliefs about the laws to ascertain which counterfactual conditionals are true, it is problematic for us to turn around and appeal to the truth or falsehood of various counterfactuals in ascertaining how laws differ from accidents in their relation to counterfactuals.

However, I am not trying to discover whether some fact is a law by consulting various counterfactuals that I know to be true only by already knowing whether that fact is a law. Rather, various truths are laws (we believe), and various counterfactual conditionals hold (we believe), and I am trying to figure out the relation between these two sets of facts. Since these facts are closely related, it is entirely to be expected that we will sometimes consult our beliefs about which truths are laws in order to arrive at our beliefs about which counterfactuals are true, just as we may sometimes use our beliefs about various counterfactuals to arrive at beliefs about the laws. ("It can't be a law that every family on my block has exactly two children, because the Jones family could have moved onto our block without violating any law of nature, and they would then still have had three children.") We have been looking at various proposals regarding the laws' special relation to counterfactuals, and we have been testing those proposals partly by seeing whether they fit our beliefs about which truths are laws and which counterfactuals are true. These tests remain severe even if we sometimes draw upon our beliefs about the laws in order to arrive at or to justify our beliefs about counterfactuals (or vice versa).

Admittedly, it would be hugely problematic if we took the laws as helping to *make* certain counterfactuals true while also taking the truth of those counterfactuals as helping to *make* certain facts qualify as laws. But I have not done that. In this chapter, I am concerned only with *identifying* the special relation between laws and counterfactuals, not with figuring out *why* this relation holds: whether laws are responsible for counterfactuals, or counterfactuals are responsible for laws, or neither is responsible for the other. We will grapple with those questions in chapters 2 and 4.

1.8. No Nonmaximal Set Containing
Accidents Possesses Sub-nomic Stability

Let's see an argument for that bold assertion!

Take a set of sub-nomic truths containing every sub-nomic logical consequence of its members and including the fact that all gold cubes are smaller than a cubic mile. What would it take for this set to be sub-nomically stable (or just "stable," for short)? As we have seen, it is not the case (in every conversational context) that all of the set's members would still have been true had Bill Gates wanted to have a large (exceeding one cubic mile) gold cube built. How can the set be stable despite failing to be preserved under this counterfactual supposition? Stability requires the set's members all to be invariant under every sub-nomic counterfactual supposition that is logically consistent with them all (taken together). The only way for this set to be stable, despite failing to be preserved under the supposition "Bill Gates wants to have a large gold cube built," is for that supposition to be logically inconsistent with the set. To be stable, then, the set's members must logically entail "Bill Gates never wants to have a large gold cube built," so that the supposition that Bill Gates wants to have a large gold cube built is logically inconsistent with the set's members. Since the set contains every sub-nomic logical consequence of its members, the set must contain the fact that Bill Gates never wants to have a large gold cube built.

However, the set's stability is not yet assured. Presumably, had Melinda Gates wanted to own a large gold cube, then Bill (Melinda's husband) would have wanted one built. Hence, a member of the set ("Bill Gates never wants to have a large gold cube built") is not preserved under the supposition that Melinda Gates wants to own a large gold cube. How can the set be stable despite failing to be preserved under this supposition? The argument we gave a moment ago applies here as well. If the set includes the fact that Melinda Gates never wants to own a large gold cube, then the supposition that she wants to own one is logically inconsistent with the set, and so the set's failure to be preserved under this supposition does not preclude its stability. Therefore, if a stable set includes the fact that Bill Gates never wants to have a large gold cube built, then it must also include the fact that Melinda Gates never wants to own one.

With each additional claim p that is admitted into the set in order to keep the set's behavior under some supposition ~p from rendering it unstable, we must worry about the suppositions ~q under which p is not preserved. To prevent the set's behavior under those suppositions from rendering it unstable, further claims q must be admitted into the set. The process snowballs until the set contains every sub-nomic truth. Thus, no nonmaximal set containing accidents possesses stability.

For example, suppose the set omits the accident that all of the apples on my tree are ripe. Then the following counterfactual supposition is logically consistent with the set: had either some gold cube exceeded one cubic mile or some apple on my tree not been ripe. Under this supposition, there is no reason why the generalization about gold cubes (which is in the set) should take priority in every conversational context over the apple generalization (which we have supposed not to be in the set). So it is not the case that in every context, the gold cubes generalization is preserved under this counterfactual supposition. Hence, to be stable, the set must include the apple generalization (thereby rendering the supposition logically inconsistent with the set). Therefore, if the set is stable and includes one accident, then it must include every accident.

Let's confirm this conclusion by looking at another example. Return to the accidental truth g: whenever the gas pedal of a certain car is depressed by x inches and the car is on a dry, flat road, then the car's acceleration is given by $a(x)$. Had the pedal on a certain occasion been depressed a bit farther, then g would still have held. However, a set containing g is unstable unless it also includes a description of the car's engine, since g might not still have held had the engine contained six cylinders instead of four. With a description of the car's engine in the set, the set's failure to be preserved under "Had the engine contained six cylinders instead of four" does not compromise the set's stability. But now to be stable, the set must also include a description of the engine factory, since had the factory been different, the engine might have been different. By packing more and more into the set, will we ever achieve stability before the set contains every sub-nomic fact?

I do not think so. Take a logically closed set containing g but omitting the fact that Jones is not wearing an orange shirt. Now consider what the world would have been like, had either ~g held or Jones been wearing an orange shirt. Would g still have held? In every conversational context? Certainly not! In at least some contexts, g might still have held, but Jones might just as well still not have been wearing an orange shirt.[33] In those contexts, it is *not* the case that a given truth (g) *within* the set would still have held, had either it or an arbitrary truth *outside* of the set been false. That is enough to make the set unstable. To disarm this threat to the set's stability, we must ensure that the set contains the fact that Jones is not wearing an orange shirt; the threatening supposition ($\sim g$ or Jones is wearing an orange shirt) is then logically inconsistent with the set, and so to be stable, the set has no need to be preserved under that supposition. But if a set containing g is rendered unstable by omitting even an arbitrary sub-nomic fact, then the set is unstable if it fails to include *every* sub-nomic fact.

The same sort of argument could be made regarding any set Γ of sub-nomic truths containing every sub-nomic logical consequence of its members and *some* accidents but not *all* of them. There exist two intuitively unrelated accidental truths, p and q, where Γ includes p but omits q. For Γ possibly to be stable, each member of Γ, such as p, must be invariant (in every conversational context) under the counterfactual supposition that either $\sim p$ or $\sim q$, since this supposition is logically consistent with Γ. (If ($\sim p$ or $\sim q$) were inconsistent with Γ, then since Γ contains every sub-nomic logical consequence of its members, Γ would have to contain $\sim(\sim p$ or $\sim q)$, that is, ($p \& q$), and since (once again) Γ contains every sub-nomic logical consequence of its members, Γ would have to contain q, which Γ was stipulated as omitting.) But it is not the case that ($\sim p$ or $\sim q$) $\square \rightarrow p$ holds in every context. In picturesque terms, the supposition ($\sim p$ or $\sim q$) pits p against q, as far as remaining true is concerned. They cannot both be preserved under this supposition; at least one must go. With p and q utterly unrelated and neither a law, it is not the case that p takes priority over q in every context, no matter which facts are salient there. Hence, Γ is unstable. Although the set has picked out a comfortable range of counterfactual suppositions, the set is not invariant under all of these suppositions.

The above argument made no appeal to nested counterfactuals, although the definition of "sub-nomic stability" requires that each member m of a stable set be preserved even under arbitrarily many nested suppositions, so that $p \square \rightarrow m$, $q \square \rightarrow (p \square \rightarrow m)$, $r \square \rightarrow (q \square \rightarrow (p \square \rightarrow m))$, . . . all hold. These nested counterfactuals may seem remote from actual scientific practice. But in fact, scientists routinely employ nested counterfactuals ("Had the chamber been completely evacuated, then had a few CO_2 molecules been present, they would have had a long mean free path"; "Had gravity declined with the cube of the distance, then a solar system, had it begun with many planets, would not long have so remained; planets would have soon escaped or spiraled into the sun"). Although the argument I just gave shows how difficult it would be, I suppose it is barely possible for there to be a nonmaximal set of sub-nomic truths containing an accident where (in every context) each member m would still have held under any sub-nomic counterfactual supposition p that is logically consistent with the set. But if there is such a set, its invariance is just a fluke. That is, its invariance is not likewise invariant. (Or if it is, then *that* is just a fluke: *its* invariance is not likewise invariant. (Or....)) In other words, although every counterfactual $p \;\Box\!\!\rightarrow\; m$ requisite for stability may hold, there is some q that is logically consistent with the set where $q \;\Box\!\!\rightarrow\; (p \;\Box\!\!\rightarrow\; m)$ fails (or one of the further nested counterfactuals fails). To ensure that no nonmaximal set containing accidents manages to possess sub-nomic stability, we need nested counterfactuals in the definition of "sub-nomic stability."

The argument that I have just given against the stability of any nonmaximal set Γ containing accidents cannot be used against Λ's stability. Although context wields great influence over counterfactuals, there is a limit to its influence: in no context does an accident q take priority over a law p under the counterfactual supposition ($\sim p$ or $\sim q$). In note 29, I look at several sorts of cases that initially might appear to violate Λ's sub-nomic stability, and I argue that all of them are best understood without denying Λ's stability. In addition to those arguments, here is a general way of thinking about why context cannot enable an accident q to take priority over a law p under ($\sim p$ or $\sim q$).

Consider any ordinary counterfactual conditional in a context where it is true. For example,

Had my family gone out to dinner last night, we would have gone to an ethnic restaurant in North Carolina.

(We live in Chapel Hill, North Carolina; we ate at home last night; when we eat out, we tend to visit local ethnic restaurants.) Let us gradually turn the counterfactual supposition into a disjunction ($\sim p$ or $\sim q$). In the same context, the following conditional is true:

Had my family gone out to dinner last night, we might have gone to Chinese Noodle Restaurant in Chapel Hill, but we would not have gone to McDonald's in Istanbul.

Hence (in the same context),

Had my family eaten dinner last night either at Chinese Noodle Restaurant in Chapel Hill or at McDonald's in Istanbul, we would have gone to Chinese Noodle and not to Istanbul.

In this context, the fact that we did not go to Istanbul last night for dinner takes priority over the fact that we did not go to Chinese Noodle for dinner. Now consider a scenario even more remote than our going to McDonald's in Istanbul last night for dinner: our breaking some law of nature last night. If our going to Chinese Noodle is "closer" in the given context than our dining at McDonald's in Istanbul, then it is "closer" than something even more outlandish than our dining at McDonald's in Istanbul: our violating the laws. How does that qualify as more outlandish?[34] There is a variety of possibility (namely, natural possibility) such that our dining last night at Chinese Noodle was possible but our violating the laws was impossible—and anything possible is "closer" than everything impossible. In other words, it is possible (naturally) for my family to eat dinner at Chinese Noodle or to break the law of gravity—and whatever would have happened, under some possible circumstance, must qualify as possible. So in the given context, this counterfactual is true:

Had my family either eaten dinner last night at Chinese Noodle Restaurant in Chapel Hill or broken the law of gravity, we would have eaten dinner last night at Chinese Noodle and not broken the law of gravity.

This argument was given for an arbitrary context where all of the above suppositions can be entertained; the same kind of argument could be given for any context. So under a supposition ($\sim p$ or $\sim q$) that pits a law's preservation against an accident's, the law takes priority in any context—since context is powerless to override the principle that anything possible is nearer than everything impossible, and there is a species of necessity associated with the natural laws.

These ideas about necessity and possibility will play prominent parts in the next chapter, where I will defend them further.

1.9. How Two Sub-nomically Stable Sets Must Be Related: Multiple Strata of Natural Laws

I have suggested that Λ is a sub-nomically stable set and that no non-maximal set of sub-nomic facts containing an accident is sub-nomically stable. Are there any other sub-nomically stable sets? There is at least one: the set containing exactly the broadly logical truths that are sub-nomic. For example, 3 would still have failed to divide 23 evenly even if I had been wearing an orange shirt, or even if gravity had declined with the cube of the distance—indeed, under any counterfactual supposition that is logically consistent with the broadly logical truths.

Furthermore, for any two sub-nomically stable sets, one must be a proper subset of the other, so the sub-nomically stable sets must fall into a natural hierarchy. Here is the proof (see fig. 1.2).

Suppose (for the sake of *reductio*) that Γ (gamma) and Σ (sigma) are both sub-nomically stable sets, t is a member of Γ but not of Σ, and s is a member of Σ but not of Γ.

Let's start with Γ. The claim ($\sim s$ or $\sim t$) is logically consistent with Γ. (Since Γ is stable, Γ contains every sub-nomic logical consequence of its members, so since Γ does not contain s, it follows that Γ does not entail s, and so $\sim s$ is logically consistent with Γ, and hence ($\sim s$ or $\sim t$) is, too.)

Since Γ is sub-nomically stable, every member of Γ would still have been true, had ($\sim s$ or $\sim t$) been the case.

In particular, t would still have been true.

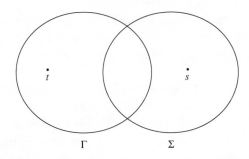

Figure 1.2   Sets Γ and Σ with their members t and s, respectively.

Thus t & ($\sim s$ or $\sim t$) would have held, had ($\sim s$ or $\sim t$).

Hence, ($\sim s$ or $\sim t$) $\square\rightarrow \sim s$.[35]

Now let's work from the Σ side. Since ($\sim s$ or $\sim t$) is logically consistent with Σ, and Σ is sub-nomically stable, all of Σ's members are preserved under the supposition that ($\sim s$ or $\sim t$). It is not the case, for any of Σ's members, that its negation would (or even might) have held, had ($\sim s$ or $\sim t$).

Take s in particular: $\sim$ (($\sim s$ or $\sim t$) $\square\rightarrow \sim s$).

But this result contradicts our earlier conclusion that ($\sim s$ or $\sim t$) $\square\rightarrow \sim s$.

The promised *reductio* has been achieved: we have shown that for any two sub-nomically stable sets, one must be a proper subset of the other.[36] This result can guide our search for other sub-nomically stable sets besides Λ and the set of sub-nomic, broadly logical truths. For instance, this result (together with the stability of the set of sub-nomic, broadly logical truths) entails that any stable set containing some sub-nomic, broadly logical truths, along with some sub-nomic truths that are not broadly logical truths, must contain every sub-nomic, broadly logical truth. But any sub-nomically stable set must contain every sub-nomic, narrowly logical truth (since those truths are logical consequences of its members), and any narrowly logical truth is a broadly logical truth. Hence, any stable set containing some sub-nomic truths that are not broadly logical truths must contain every sub-nomic, broadly logical truth. Furthermore, as we saw in the previous section, no nonmaximal set containing an accident is stable. Thus, no nonmaximal superset of Λ is stable. Therefore, to find further promising candidates for stability, let's look among proper subsets of Λ containing all of the sub-nomic, broadly logical truths.

Many of these sets clearly lack sub-nomic stability. Consider Coulomb's law: between any two point bodies that have for a long while been at rest at any positions r_1 and r_2 ($r_1 \neq r_2$) while carrying any electric charges q_1 and q_2, there is a mutual electrostatic repulsion $F = q_1 q_2 / |r_1 - r_2|^2$. Now consider a logical consequence of Coulomb's law, namely, that the above Coulombic regularity holds at all times *after today*. Take the proper subset of Λ containing exactly this restricted Coulombic regularity, the broadly logical sub-nomic truths, and their sub-nomic logical consequences. Is this set stable? Consider this counterfactual supposition: Had Coulomb's law been violated sometime *before* today. This supposition is logically consistent with the restricted Coulombic regularity, since that regularity concerns events only after today. Therefore, the chosen set is sub-nomically stable only if its members are all preserved under this supposition. But they are not. Had Coulomb's law been violated sometime before today, then Coulomb's "law" would not have been a law and so would not have been around to mandate the restricted Coulombic regularity. With Coulomb's law out of the way, there would have been nothing to keep the course of events after today from violating Coulomb's law. Therefore, it might well have been violated after today—just as energy might well have failed to be conserved, had there been no law making energy conservation compulsory.

However, *some* proper subsets of Λ containing all of the broadly logical sub-nomic truths *are* plausibly sub-nomically stable. Take the fundamental law of dynamics, which governs the relation between the forces on a body and the body's motion. At one time, this law was believed to be Newton's second law of motion $F = ma$, relating the net force F on a body to its mass m and acceleration a. In 1830, George Biddell Airy used Newton's second law of motion to figure out how bodies would have behaved had they been subjected to various weird hypothetical kinds of forces.[37] His investigation presupposes that Newton's second law of motion would still have held, even if the force laws had been different. Similarly, Paul Ehrenfest in 1917 famously showed that had gravity been an inverse-cube force or fallen off with distance at any greater rate, then planets would eventually have collided with the sun or escaped from the sun's gravity.[38] Ehrenfest's argument also presumes that Newton's second law of motion would still have held, had gravity obeyed a different force law.

Plausibly, the fundamental dynamical law would still have held, had the world been populated by different kinds of forces or different kinds of fundamental particles, or had the strengths of those forces or the characteristic properties of those particles been different. Had the electrostatic force been half as strong, or light speed been half as fast, or the electron's charge been half as great, then the fundamental dynamical law would have been no different. Any additional kinds of forces and particles, had they existed, would have obeyed the same fundamental dynamical law as the actual kinds do. The fundamental dynamical law would still have held, had there been charged leptons other than muons, electrons, and taus—the actual species of charged leptons. (This counterfactual supposition violates the "closure law" belonging to Λ that all charged leptons are muons or electrons or taus. Closure laws will arise again in chapter 3.)

Likewise, consider the law of the composition of forces. It specifies how various component forces, of whatever kind, add (via the "parallelogram of forces") to yield a total resultant force—the "net force" appearing in Newton's second law of motion. The law of the composition of forces would still have held, even if the world had been populated by different kinds of forces.[39] By the same token, it is commonly suggested that the conservation laws transcend the particularities of the force laws. Even if there had been different kinds of forces (or, say, the electrostatic force had been twice as strong), momentum and energy would still have been conserved. Regarding energy conservation, momentum conservation, and so forth, Eugene Wigner writes: "[I]t is clear that their validity transcends that of any special theory—gravitational, electromagnetic, etc.—which are only loosely connected...."[40]

Likewise, consider the Lorentz transformations, which are central to Einstein's special theory of relativity, entailing such famous relativistic results as time dilation and length contraction. They specify how an event's space-time coordinates in one reference frame in a certain family relate to its coordinates in that family's other frames. In his first relativity paper in 1905, Einstein derived these transformations by using the principle that light travels at the same rate in one of these frames whatever the motion of its source. However, as Einstein later wrote,[41] this derivation is misleading because features of light, a particular inhabitant of space-time, are not responsible for the coordinate transformations. The widespread recognition that the transformations lie deeper than the particular kinds of fundamental forces or particles there happen to be has led to a long tradition, beginning as early as 1909, of deriving the Lorentz transformations without appealing to any details of electromagnetism or any other force.[42] It is only because scientists believe that the Lorentz transformations transcend the fundamental force laws that they believe that were there additional fundamental force laws or had the fundamental force laws been different, the Lorentz transformations would still have held. As Roger Penrose says, if it is just "a 'fluke'" that certain dynamical laws exhibit Lorentz covariance, then "[t]here is no need to believe that this fluke should continue to hold when additional ingredients of physics are" discovered. But physicists generally do regard special relativity as prior to the force laws.[43] For instance, physicists commonly assert that had the force laws been different so that photons, gravitons, and other kinds of particles that actually possess zero mass instead possessed nonzero mass, the Lorentz transformations would still have held (though these particles would not have moved with the speed c that famously figures in the transformations).[44]

Thus, by using the concept of sub-nomic stability, we can cash out what it would be for the conservation laws, the parallelogram of forces, the fundamental dynamical law, and the coordinate transformations to rise above the specific kinds of forces there happen to be. It would be for a set containing laws such as these (and the sub-nomic, broadly logical truths), but omitting the force laws, to possess sub-nomic stability. Here we have a plausible candidate for a stable proper subset of Λ. In short, there appears to be a hierarchy of sub-nomically stable sets that includes at least these members (see fig. 1.3).

The pyramid in figure 1.3 suggests that there are at least two "strata" or "levels" of natural law. Any metaphysical account of what natural laws are should leave room for laws to come in multiple strata.

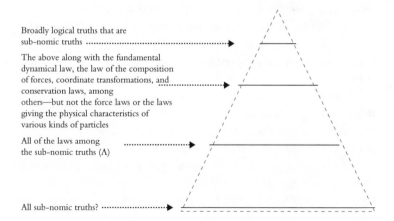

Broadly logical truths that are
sub-nomic truths ⋯⋯⋯⋯⋯⋯⋯⋯⋯⋯⋯⋯⋯➤

The above along with the fundamental
dynamical law, the law of the composition
of forces, coordinate transformations, and ⋯⋯⋯⋯⋯➤
conservation laws, among
others—but not the force laws or the laws
giving the physical characteristics of
various kinds of particles

All of the laws among   ⋯⋯⋯⋯⋯⋯➤
the sub-nomic truths (Λ)

All sub-nomic truths? ⋯⋯⋯⋯⋯⋯➤

Figure 1.3   Some (though perhaps not all) good candidates for sub-nomically stable sets. This pyramid is *not* the hierarchy depicted in figure 1.1, which placed sub-nomic facts at the bottom, the laws governing *them* (first-order laws) one rung higher, the laws governing them (meta-laws) one rung higher, and so forth. Unlike that earlier hierarchy, every rung of this pyramid contains exclusively sub-nomic facts.

In later chapters, I shall impose this criterion of adequacy on familiar philosophical accounts of natural law and say more about the significance of the laws' coming in multiple strata.

1.10. Why the Laws Would Still Have Been Laws

I have argued that it is a law that m if and only if m belongs to at least one nonmaximal sub-nomically stable set (or equivalently—since the sub-nomically stable sets form a pyramid—if and only if m belongs to the largest nonmaximal sub-nomically stable set). I have also suggested that this biconditional captures the special relation in which laws stand to counterfactuals. It is a consequence of what lawhood is, not a peculiarity of what the actual laws are in fact like. If these suggestions are correct, then a philosophical analysis of what a natural law is should account for this relation between lawhood and stability. I will take up that challenge later in this book. For now, I shall note only that any analysis that succeeds in explaining why stability and lawhood are so related will automatically have a further payoff: it will thereby explain why the (first-order) laws would still have been *laws* under any sub-nomic counterfactual supposition p that is logically consistent with the laws. Here the nested counterfactuals in the definition of sub-nomic stability come into play.

Suppose that m is a member of Γ, a sub-nomically stable set, and that each of $q, r, s \ldots$ is individually logically consistent with Γ. Then Γ's stability ensures that $\sim (q \diamond\!\!\rightarrow \sim m)$, $\sim (q \diamond\!\!\rightarrow (r \diamond\!\!\rightarrow \sim m))$, $\sim (q \diamond\!\!\rightarrow (r \diamond\!\!\rightarrow (s \diamond\!\!\rightarrow \sim m))$, and so forth. One of the connections we saw (in section 1.5) between might-conditionals and would-conditionals was that $\sim (q \diamond\!\!\rightarrow \sim m)$ logically entails $(q \square\!\!\rightarrow m)$. So the counterfactuals in the above sequence respectively entail $(q \square\!\!\rightarrow m)$, $(q \square\!\!\rightarrow \sim (r \diamond\!\!\rightarrow \sim m))$, $(q \square\!\!\rightarrow \sim (r \diamond\!\!\rightarrow (s \diamond\!\!\rightarrow \sim m))$, and so forth. So had q been the case (that is, in the "closest q-world"), the following all hold: m, $\sim (r \diamond\!\!\rightarrow \sim m)$, $\sim (r \diamond\!\!\rightarrow (s \diamond\!\!\rightarrow \sim m))$, and so forth—simply each of the earlier counterfactuals with their opening $q \square\!\!\rightarrow$'s lopped off, since we are talking about what's true in the closest q-world. But this sequence supplies exactly what is needed for Γ to be sub-nomically stable in that world: m and its colleagues in Γ are all true and preserved under every counterfactual supposition that is logically consistent with Γ. Hence, if Γ is in fact sub-nomically stable, then Γ would still have been sub-nomically stable had q been the case, for any q that is logically consistent with Γ.

Therefore, for any such q, the actual laws would still have been *laws* had q been the case—if the laws under q are exactly the members of at least one set that would under q have been nonmaximal and sub-nomically stable. (Moreover, any *stratum* of laws would still have constituted a stratum of laws, had q been the case.) We thereby save the intuition that had Earth's axis of rotation been nearly aligned with its orbital plane (so that Earth was "lying on its side," as Uranus actually is), then although terrestrial seasons would have been quite different, the actual laws of nature would still have been laws—which is *why* terrestrial seasons would have been so different. The laws' collective invariance under counterfactual suppositions is no "accident"; the laws' invariance (and hence their lawhood) is itself invariant. This fact will resurface in the chapters that follow.

1.11. Conclusion: Laws Form Stable Sets

This chapter sets up everything else in the book. So let's review the main point: Λ is the largest nonmaximal sub-nomically stable set. Whereas NP uses the *laws* to pick out a range of counterfactual suppositions under which a set's invariance is to be tested, stability allows any set to pick out a comfortable range for itself. Thus, in using stability to explain how laws differ from accidents in having a special relation to counterfactuals, we avoid arbitrarily privileging the laws from the outset; we avoid specifying the laws as the truths that would still have held under every counterfactual supposition that is logically consistent with the laws. Although we could talk about the set of truths that are invariant under every supposition that is logically consistent with (say) "George Washington is the first president of the United States," these truths fail to form a stable set. By identifying the laws as the members of at least one nonmaximal stable set, we discover how a sub-nomic fact's lawhood is fixed by the sub-nomic facts *and the subjunctive facts about them*. The laws' stability turns out to account not only for the sharp distinction between laws and accidents, but also for the fact that the laws would still have been *laws* had *q* been the case, for any *q* that is logically consistent with the laws.

I began this chapter by suggesting that laws are set apart from accidents by their *necessity* (of a certain kind). Let's now see how the notion of stability helps us to understand the laws' characteristic necessity ("natural necessity"). On to chapter 2!

2

Natural Necessity

2.1. What It Would Take to Understand Natural Necessity

Chapter 1 began with the thought that laws are set apart from accidents by their necessity. The laws' characteristic variety of necessity is often termed "*natural* necessity"[1] to distinguish it from other putative species of necessity, such as:

> (*Narrowly*) *logical* necessity (as possessed by the fact that either all emeralds are green or some emerald is not green, for example)[2]
>
> *Conceptual* necessity (all sisters are female)[3]
>
> *Mathematical* necessity (there is no largest prime number)
>
> *Metaphysical* necessity (red is a color)
>
> *Moral* necessity (one ought not torture babies to death for fun)[4]
>
> *Broadly logical* necessity (as possessed by a truth in any of these categories)

Our principal goal in this chapter is to understand what natural necessity is.[5]

An adequate account of natural necessity must explain how the laws manage to be genuinely necessary but nevertheless contingent:

- "Genuinely necessary": Whatever natural necessity is, it must deserve the name by truly being a variety of *necessity*—a species of the same genus as the variety (or varieties) of necessity possessed by broadly logical truths.

- "Contingent": Although the laws are necessary, they are not *as necessary as* the truths possessing logical, conceptual, mathematical, metaphysical, or moral necessity. Natural necessity is a weaker variety of necessity.

In this chapter, I shall argue that by using the concept of "sub-nomic stability" (the kind of invariance under counterfactual suppositions that I elaborated in chapter 1), we can give an adequate account of natural necessity. In particular, we can explain what natural necessity has in common with other varieties of necessity by virtue of which they all qualify as varieties of *necessity*. We can also explain why no distinctive variety of necessity is possessed by the members of an arbitrary set of truths. In addition, we can identify what makes one variety of necessity "stronger" than another. Indeed, we can explain why all varieties of necessity (insofar as they pertain to sub-nomic truths) have characteristic places in a single, well-defined ordering from strongest to weakest.

This chapter's main thesis will be that for each variety of necessity, the sub-nomic truths possessing it form a sub-nomically stable set, and for each sub-nomically stable set (except the set of all sub-nomic truths, if it be stable), there is a variety of necessity where the sub-nomic truths so necessary are exactly the set's members. In short, there is a correspondence between the species of necessity and the nonmaximal sets possessing sub-nomic stability.

The laws' status as contingent yet necessary has seemed so paradoxical that some philosophers have apparently felt compelled to deny one of these two features in order to embrace the other. As I see it, David Lewis's "Best System Account"[6] respects the laws' contingency but fails to do justice to their necessity. Laws in Lewis's picture retain nothing that we would ordinarily recognize as genuine necessity. Conversely, "scientific essentialism"[7] respects the laws' necessity but not their contingency. By identifying natural necessity with metaphysical necessity, essentialism fails to find a place for the laws somewhere *between* the metaphysical necessities and the accidents.

I stubbornly demand that an account of laws give me everything that I originally wanted: the laws' necessity *and* their contingency!

2.2. The Euthyphro Question

In Plato's *Euthyphro*, Socrates famously asks whether good actions are good by virtue of having the gods' approval, or whether good actions have the gods' approval because they are good. If the gods' approval is what *makes* good actions good, then (as Leibniz says) "one destroys . . . all the love of God and all His glory; for why praise Him for what he has done, if He would be equally praiseworthy in doing the contrary?"[8]

A version of the Euthyphro question is suggested by our opening thought: that the laws are the truths possessing natural necessity. Are the laws necessary by virtue of being laws, or are they laws by virtue of being necessary? Is their necessity what makes them laws, or does their status as laws bestow necessity upon them? It seems to me that intuitively, their necessity is what makes them laws. If, on the contrary, their necessity arises merely ex officio from their lawhood, then praising the laws for their necessity would be as empty as praising the gods for loving the good, if being good is nothing more than being loved by the gods.

Here is one standard definition of natural necessity:

p is "naturally necessary" if and only if p logically follows from the m's (taken together) where it is a law that m.

A slightly different definition is sometimes used instead:

p is "naturally necessary" if and only if p logically follows from the truths (taken together) of the form "It is a law that m" and "It is not a law that m."

On my view, these two definitions identify exactly the same sub-nomic facts as naturally necessary, and both are correct in this regard.[9] But each definition tends to be presented as setting forth what natural necessity *consists of*. They are presented as *reductive* definitions: as reducing p's natural necessity to a fact about the laws. If this is correct—that is, if lawhood is ontologically prior to natural necessity—then a fact's natural necessity cannot be responsible for its lawhood, on pain of circularity. What *makes* a law out of p cannot be p's natural necessity if p's natural necessity, in turn, is nothing more than p's following from the laws. Thus, the standard reductive definitions of natural necessity presuppose an answer to the Euthyphro question: that the laws' natural necessity is *not* what fundamentally sets them apart from the accidents. Rather, being naturally necessary is just a perk (a fringe benefit) of being a law.

However, it is then an empty compliment to praise the laws for being "naturally necessary." Of course, it is no small achievement for a given p to be naturally necessary—that is, to be a matter of law. But it is a trifling fact that the laws (whatever they are) are naturally necessary. That the laws follow from the laws is not much for them to brag about, since any claims follow from themselves. If following from *the laws in particular* is supposed to be worth bragging about, then why is that? What is so special about the laws that the distinction between what follows from them and what doesn't is significant? After all, we could stipulate the "Washington necessities" to be the truths that follow from the facts about whether or not George Washington was a president of the United States and (if so) which number president he was. But we would not ordinarily regard some p's status as a "Washington necessity" as carrying any cosmic importance or as making p genuinely necessary.

The gods are praiseworthy for loving the good only if the goodness of what the gods love is prior to the gods' loving it. By the same token, when Captain Kirk insists that Chief Engineer Scott give him warp power now and Scotty protests, "I cahnuh change the laws o' physics, Cap'n,"[10] Scotty is not blameworthy for failing to obey the Captain's order. But it is difficult to see why Scotty has a good excuse if the laws possess their "necessity" only ex officio. Admittedly, the laws cannot be violated without breaking the laws. But why does that mean that the laws cannot be violated?

Here we have analogues to the problems of triviality, circularity, and arbitrary privilege that we encountered when we tried (in chapter 1) to use NP to capture the laws' special relation to counterfactuals. This parallel is no coincidence. Just as NP allows the laws to fix the relevant range of counterfactual suppositions to suit themselves, so the above definitions of "natural necessity" give a special role to the laws. This parallel will continue. In chapter 1, I suggested that we avoid arbitrarily privileging the laws by granting to every set of truths the right to pick out for itself a comfortable range of counterfactual suppositions—giving every set the opportunity to play the laws' role in NP. Thus we arrived at the concept of "sub-nomic stability." By making an analogous move, we will shortly arrive at an account of necessity.

If being "naturally necessary" were nothing more than being a consequence of the laws, then for the laws to boast merely of their natural necessity would leave us unimpressed. Unless there were *already* something special about being a law, there would be nothing special about being naturally necessary. Of course, there *is* already something special about the laws according to philosophers who define "natural necessity" in terms of lawhood. Those philosophers may disagree about what lawhood is, but on each of their accounts, being a law involves something metaphysically prior to being naturally necessary. However, in so replying to the Euthyphro question, any such account is potentially open to a powerful objection: whatever is supposed to make various facts qualify as laws, those "lawmakers" seem unable to make those facts genuinely *necessary*. Over the course of the next three sections, I will work toward explaining how this objection can be offered against two justly celebrated accounts of natural law.

2.3. David Lewis's "Best System Account"

According to Lewis, two possible worlds are exactly alike regarding their laws if they are exactly alike regarding where in space and when in history there occur instantiations of various properties of a certain elite sort. That is, the facts about which truths are laws and which are not "supervene" on the facts about space-time's geometry and the global spatiotemporal distribution (or "mosaic") of elite-property instantiations.[11] A property is "elite" if and only if it meets these conditions:

- It is perfectly natural—unlike, for instance, the property of being electrically-charged-or-greater-than-three-inches-long. Things that share a perfectly natural property resemble one another in some genuine respect.

- It is categorical—that is, "Humean": its relations to any modalities, propensities, chances, laws, counterfactuals, dispositions, and so forth are not essential to it. The instantiation of various elite properties here now imposes through logic alone no constraints at all upon which other elite properties are instantiated anywhere (or anywhen). In contrast, that this bottle would shatter, were it struck, is not a categorical property of the bottle, since its possessing this property together with its now being struck logically demands that it shatter.[12]

- It is "qualitative" in that it does not involve a "haecceity" (a "being-this-thing-ness"): the property that (some philosophers say) a given thing possesses of being the particular thing it is.

- It is possessed intrinsically by a space-time point or occupant thereof.[13] Roughly, a thing possesses an "intrinsic" property by virtue of what it is like in itself, regardless of what the universe's other inhabitants are like, whereas possessing some nonintrinsic property involves standing in a certain relation to something else. For example, the property of being taller than Jones is nonintrinsic in that whether Smith possesses this property depends not only on Smith's height, but also on Jones's.

Also supervening on the Humean mosaic (according to Lewis) are facts about single-case objective chances, such as this atom's having a 50% chance of undergoing radioactive decay in the next 138.39 days.

Lewis's analysis of lawhood has us consider the various deductive systems consisting of (i) some (but perhaps not all of the) truths regarding instantiations of elite properties and (ii) claims ascribing some chance to various elite properties' later being instantiated if certain elite properties have already been instantiated (for example, the claim that for any time T, any polonium-210 atom at T has a 50% chance of decaying in the 138.39 seconds after T). These rivals[14] compete for the coveted title of "Best System" along three dimensions:

1. Informativeness (in excluding or ascribing chances to possible arrangements of elite-property instantiations—so "All emeralds are green" counts as more informative than "All emeralds before the year 3000 are green"),

2. Simplicity (for instance, in the number of axioms in the deductive system and the order of polynomials therein, when the system's members are expressed in terms of perfectly natural properties, space-time relations, and chances), and

3. Fit (where a system fits the actual world better insofar as it assigns a higher probability to the actual course of elite-property instantiations—so that if, over the course of history, there are many atoms of polonium-210 and about 50% of them undergo radioactive decay within 138.39 seconds of coming into existence, then "All atoms of ^{210}Po have a half-life of 138.39 seconds" achieves a better fit than "All atoms of ^{210}Po have a half-life of 20 seconds").

These three criteria stand in some tension. Greater informativeness can be achieved by adding claims to the system, which often (though not always) diminishes the system's simplicity. If various atoms of ^{210}Po decay after having existed for various periods of time, then a system specifying the exact span of time during which each of these atoms exists is very informative but very complicated, whereas a system specifying that each of them had a 50% chance of decaying within 138.39 seconds of coming into existence is much less informative but much simpler and may fit the world reasonably well.

Perhaps some single system is by far the best on balance at meeting these three criteria. Perhaps which system "wins" the competition is relatively insensitive to any arbitrary features of our standard of simplicity or our rate of exchange among the three criteria.[15] In that case, Lewis says, the laws of nature are the contingent generalizations belonging to the Best System. Moreover, the facts about the chances (at a given moment) of various elite properties being instantiated at later moments are whatever the Best System (together with the history of elite-property instantiations through that moment) entails them to be.

2.4. Lewis's Account and the Laws' Supervenience

Lewis's account has the virtue (according to many philosophers) of using only Humean resources to distinguish laws from accidents.[16] It also nicely accommodates "uninstantiated laws." Take Coulomb's law, which specifies the electrostatic force between any two point charges that have long been at rest. Take the Best System, and replace Coulomb's law in it by a generalization agreeing with Coulomb's law except for any case involving a point charge of exactly 1.234 statcoulombs located exactly 5 centimeters from a point charge of exactly 6.789 statcoulombs. For that particular case, let the replacement predict twice the electrostatic force that Coulomb's law predicts. Suppose there never exists such a pair of charges, so the replacement is true, just like Coulomb's law. However, it is more complicated than Coulomb's law, since it accords separate treatment to this particular case. Hence, the Best System is simpler than (and just as informative and fit as) this alternative, so it beats this alternative in the competition for Best System. On Lewis's account, then, the laws include Coulomb's law, replete with uninstantiated cases.

That the laws supervene on the Humean mosaic appears to render them epistemically more accessible and metaphysically less mysterious than they would otherwise be. However, Lewis's picture seems to me deeply counterintuitive and radically at odds with scientific practice. The laws have often been characterized metaphorically as the rules of the game that is being played by the entities populating the universe (particles, fields, or whatever).[17] The laws have also been characterized as the "software of the universe," directing the functioning of the hardware (particles, fields, or whatever).[18] These metaphors suggest that contrary to Lewis's account, the laws fail to supervene on the Humean mosaic. After all, the rules of a game fail to supervene on the moves that are actually made on one occasion (or every occasion) when the game is played. If neither player ever castles in the course of a game like chess, for instance, then the moves they make leave undetermined whether castling was permitted.[19] Likewise, if the values for some variable N that we input to some computer program never exceed 500, then how the program would have operated on higher values for N is not determined by the routines it performs on the inputs it actually receives.

The laws seem not to supervene on the Humean mosaic. Consider two possible worlds involving the same Humean mosaic (where all F's are G, let's say), but whereas in one world it is a law that all F's are G, this regularity is accidental in the other world; there it is a law that all F's have 99.99% chance of being G. Or try a more radical intuition–pump: suppose there had been nothing in the entire history of the universe except a single, lonely electron moving uniformly forever.[20] Lewis's "Best System" for such a world presumably includes "At all times, there exists only one body," since that very simple truth supplies a great deal of information about the world's Humean mosaic. However, although there is a possible lone-electron world where the one-body generalization is a law, there is apparently also a possible lone-electron world where the one-body generalization is an accident. Indeed, it seems perfectly possible for a lone-electron world to have exactly the actual laws as its laws. Admittedly, many of these laws have nothing to explain in that world (just as the subroutine for N>500 is never called upon by the computer program when every input for N is below 500). But presumably even in the actual world, plenty of laws never explain anything (such as, we supposed earlier, the law specifying the electrostatic force on a point charge of 1.234 statcoulombs located 5 centimeters from a point charge of 6.789 statcoulombs).[21]

According to the intuitions behind NP (which launched our journey toward "sub-nomic stability" in chapter 1), the actual laws would all still have been laws had there been nothing but a single lonely electron, just as they would all still have been laws had Jones missed his bus to work this morning. Although there are lone-electron worlds with different laws, the "closest" lone-electron worlds (that is, the ones evoked by the counterfactual antecedent "Had there been nothing but a single lonely electron") have the same laws as the actual world. Let's dig into this.

Here is a principle that we encountered in chapter 1: the laws would still have been true under any counterfactual supposition that is logically consistent with the laws. More precisely: for each m where it is a law that m (and in every conversational context), $p \ \square\rightarrow m$ for any p where $\Lambda\cup\{p\}$ is logically consistent. ("Λ" is the set of truths m where it is a law that m; I continue chapter 1's policy of reserving lower-case letters for sub-nomic truths.) This principle is logically compatible with Lewis's Best System Account of law; the criterion for the "closest possible world" where p obtains (if p is logically consistent with the actual laws) could demand that the actual laws remain true in the closest p-world.[22] For example, by such a criterion, had there been nothing but a single lonely electron, the actual laws would all still have been true, though the Best System would have been different (for example, it would have included "At all times, there exists only one body"). However, in chapter 1, we went further and added *nested* counterfactuals to the above principle, requiring not only that every law m would still have been true had p been the case but also that $(q \ \square\rightarrow m)$ would still have been true had p been the case, for any sub-nomic claims p and q where $\Lambda\cup\{p\}$ is logically consistent and $\Lambda\cup\{q\}$ is logically consistent. For example, had we access to 23rd-century technology, then had we tried to accelerate a body from rest to beyond the speed of light, we would have failed.

Suppose that the Best System Account of law is combined with the principle that the laws of world W are still true in the closest p-world to W as long as p is logically consistent with W's laws. What does this combination have to say about nested counterfactuals? The Best System Account says that had there been nothing but a lone electron, it would have been a law that there exists exactly one body. Just as "All unicorns are white" and "All unicorns are black" both follow from

"There are no unicorns," so any claim of the form "All pairs of bodies are..." (such as Coulomb's law) follows logically from "There exists exactly one body." Since logical consequences of laws are laws, Coulomb's law and any alternative to it regarding the force between point charges would (on this view) all have been laws, had there been nothing but a lone electron. Now consider what counterfactual conditionals would have held, had there been nothing but a lone electron. That is, consider what counterfactuals hold in the closest lone-electron world. For example, is it true there that had there been more electrons than one, then any two electrons would have repelled each other in accordance with Coulomb's law? According to the Best System Account, the supposition of more than one electron is logically inconsistent with the laws in the closest lone-electron world. Therefore, the principle mandating the laws' preservation under any supposition with which the laws are logically consistent does *not* require the truth of the nested counterfactual "Had there been nothing but a lone electron, then had there been more electrons than one, the force between any two electrons would have accorded with Coulomb's law."

This should come as no surprise. We saw in chapter 1 that by way of the nested counterfactuals in the definition of "sub-nomic stability," Λ's sub-nomic stability entails that the actual laws would not only still have been *true,* but also still have been *laws* had p been the case, for any p where $\Lambda \cup \{p\}$ is logically consistent. That is contrary to the Best System Account. If the actual laws form a sub-nomically stable set in the closest p-world (as entailed by the truth in the actual world of various nested counterfactuals), then the actual laws are laws in the closest p-world, though they may well not belong to the Best System there.

For fans of supervenience, the only consolation I can offer is that if (as I argued in chapter 1) Λ is the largest nonmaximal set that is sub-nomically stable, then the facts about which m's are laws and which are accidents supervene on *something:* the sub-nomic facts *together with* various subjunctive facts. That's a super*duper*venience!

I accept counterfactuals like "Had there been nothing but a lone electron, then had there been more electrons than one, the force between any two would have accorded with Coulomb's law."[23] Fans of Lewis's account might simply deny this counterfactual, claiming that had there been nothing but a lone electron, the laws would have been so different that it would have been false that, had there been more electrons, the force between any two would have accorded with Coulomb's law. Fans of Lewis's account might argue that intuitions about nested counterfactuals with radically false antecedents are insufficiently robust to be worth saving—or that even if these intuitions are robust, they are prompted by an incorrect picture of laws as failing to supervene on the Humean mosaic.

Of course, if we take seriously the possibility that our intuitions have been corrupted by our philosophy, then we must step back, reconsider our conclusions, and seek reflective equilibrium. Naturally, I have tried to craft an account of law that saves the phenomena I find most central to scientific practice. To a reader who rejects the nested counterfactual conditional that we have been entertaining, I can say only that an otherwise attractive account of law entailing that counterfactual will thereby support it. You will have to evaluate whether my account's other features are sufficiently attractive to overcome your objections to this counterfactual.

To accept Lewis's Best System Account along with the counterfactual "Had there been nothing but a lone electron, then had there been more electrons, the force between any two would have accorded with Coulomb's law," you might consider extending the requirement that the actual laws still be true in the closest p-world (if p is logically consistent with the actual laws) to require additionally that the actual laws still be true in the closest q-world to the closest p-world to the actual world (as long as p is logically consistent with the actual laws and q is, too). But this is an unattractive option. If the closest p-world's laws differ from the actual laws, as the Best System Account entails for the closest lone-electron world, then surely the closest p-world's laws (rather than the actual laws) should influence which q-worlds count as closest to the closest p-world.

Even the more modest requirement that the actual laws still be true in the closest p-world (if p is logically consistent with the actual laws) sits uncomfortably with the Best System Account. Take a counterfactual conditional deemed true on this view (and on mine): Had there been nothing in the universe's history except for two electrons, then Coulomb's law would still have been true. In the closest worlds evoked by that supposition, what is the scientific explanation of the truth of Coulomb's law? Why, in such a world, is Coulomb's law true?

Here is how I would answer that question: "In that world, Coulomb's law is *true* because it is a *law*. The reason in that world why all electrostatic forces accord with Coulomb's law is that Coulomb's law is a law there, just as the reason in the actual world why all electrostatic forces accord with Coulomb's law is that Coulomb's law is a law here." But the Best System Account cannot give that answer. On that account, the laws actually responsible for inter-electron forces differ from the laws that would be responsible for them if there were nothing but two electrons. But if the laws in the closest p-world differ from the actual laws, then why in the closest p-world are the actual laws nevertheless true? What is the scientific explanation (in that world) for their truth?

It seems to me that just as the counterfactual ($p \,\square\!\!\rightarrow m$) is true because it is actually a law that m, so in the closest p-world, the reason why m holds is because it is a law that m. Had the match been struck, the reason why it would have lit (namely, because it would have been a law that m) corresponds to the reason why the match would have lit, had it been struck (namely, because it is actually a law that m).[24]

I anticipated this point in chapter 1, where I said that had Earth's axis been nearly aligned with its orbital plane (so that Earth is "lying on its side"), then terrestrial seasons would have been quite different, but the actual laws of nature would still have been laws—which is *why* terrestrial seasons would have been so different: because under the actual laws, the tilt of Earth's axis makes a big difference to terrestrial seasons.

2.5. The Euthyphro Question Returns

Lewis takes the laws as arising "from below"; they are constituted by the Humean mosaic. By contrast, David Armstrong takes the laws as imposed on the Humean mosaic "from above"; the laws are facts over and above the facts they govern.[25] According to Armstrong, the laws are irreducible, contingent relations of "nomic necessitation" among universals. Laws are not regularities among property instantiations; they are relations among properties, such as the properties of being electrically charged, being near another body, and being subject to a force toward that body (for the law that opposite charges attract). Such a relation's holding among universals (that Fness nomically necessitates Gness, in the simplest case) ensures a corresponding uniformity among Humean facts (that all F's are G), and that uniformity is thus naturally necessary. The nomic-necessitation relations among universals, though contingent, would still have held had Jones missed his bus to work this morning or had there been a single lonely electron, since the relations among universals stand aloof from the world's vicissitudes.[26]

I embarked on this whirlwind tour of Lewis's and Armstrong's accounts of law after promising you (at the end of section 2.2) a general recipe for offering a potentially powerful objection against any account of law that answers the Euthyphro question ("Are the laws necessary by virtue of being laws, or are they laws by virtue of being necessary?") by deeming lawhood to be metaphysically prior to natural necessity, as both Lewis's and Armstrong's accounts do. The objection involves asking how the alleged lawmakers manage to make the laws (or their consequences) *necessary*.[27]

On Armstrong's account, it is a contingent fact that Fness nomically necessitates Gness. Hence, even if a nomic-necessitation relation must be accompanied by a regularity, there is no logical, conceptual, mathematical, or metaphysical necessity that all F's be G. How does the fact that Fness "nomically necessitates" Gness supply the fact that all F's are G with natural necessity? I cannot see why that regularity should qualify as possessing some variety of necessity merely because it accompanies a certain contingent relation's holding among universals. As Lewis has remarked, the fact that Armstrong *calls* that relation "nomic necessitation" does not mean that it "can do what it must do to deserve that name."[28] If we stipulate that "naturally necessary" means nothing more than "imposed by the nomic-necessitation relations among universals," then what we have stipulated "natural necessity" to be was unrecognizable beforehand as a variety of necessity—as belonging to the same genus as logical, mathematical, conceptual, metaphysical, and moral necessity.

Lewis's account is vulnerable to the same sort of objection. I agree with Van Fraassen: "Truth and simplicity just do not add up to necessity, as far as my intuitive reactions are concerned."[29] After all, it is a law that the total quantity of electric charge in the universe is conserved. Hence, there is some truth of the form, "The total quantity of electric charge at each moment is Q." This truth presumably belongs to the Best System: it adds great strength at very little cost in simplicity. But intuitively, it is just an accident. Had the total quantity of electric charge been 2Q, the laws would still have held.

Of course, that *p* follows from the generalizations in the Best System entails that *p* is true, since the Best System's members are perforce true.

Nevertheless, there remains an open question: "Granting that p follows from the generalizations in the Best System, why is it that p *had to* obtain?" Lewis downplays the need to explain why being a consequence of the generalizations in the Best System amounts to being necessary:

> If you're prepared to grant that theorems of the best system are rightly called laws, presumably you'll also want to say that they underlie causal explanations; that they support counterfactuals; that they are not mere coincidences; that they and their consequences are in some good sense necessary; and that they may be confirmed by their instances. If not, not. It's a standoff—spoils to the victor.[30]

But it seems perfectly appropriate to insist that the lawmakers supply something that is pretheoretically recognizable as a species of necessity.

Compare Lewis on proposed analyses of objective chance. He insists that whatever entities a philosopher construes chances to be, those entities have got to deserve in advance to be called "chances."[31] In particular, beliefs about the alleged chancemakers must constrain our opinions (on pain of irrationality) in the way that beliefs about chances do—namely, in accordance with the "Principal Principle." (It says roughly that if I believe a given chance set-up [such as a coin to be flipped] has n% chance of yielding a given outcome [such as a head] and I have no other information about what the outcome will be, then my degree of confidence in that outcome should be n%.) It seems to me that regarding proposed accounts of law, such as Lewis's, we should adopt the same principled stand Lewis takes regarding proposed accounts of chance. Generalizations in the Best System must thereby deserve to count as necessary.

That is my general recipe for causing trouble for any account that takes lawhood to be metaphysically prior to natural necessity.[32] I advocate the reverse order. Later in this chapter, I offer an account of natural necessity, and in chapter 4, I contend that laws are set apart from accidents by virtue of their natural necessity.

However, I must first address another motivation for regarding "natural necessity" as merely a title of respect due to whatever occupies the office of natural law rather than as a property, the possession of which by a sub-nomic fact is responsible for making it a law. Inspiration for this argument can also be found in Lewis's writings.

2.6. Are All Relative Necessities
Created Equal?

Lewis's and Armstrong's lawmakers are ill-equipped to make any facts naturally necessary unless being naturally necessary is nothing more than following logically from the laws. But perhaps that's all natural necessity is—and not because "natural necessity" fails to deserve the name. Perhaps *every* variety of necessity is just *relative* necessity: nothing but being a logical consequence of some or another designated class of facts (designated not by their necessity, of course). For natural necessity, those facts are the laws. It would then be appropriate for the laws' natural necessity to be no great credit to them, since for any variety of necessity, the necessity of the designated class of facts (the facts that this necessity is relative to) is trivial. If every variety of necessity (except perhaps narrowly logical necessity) is merely relative necessity, then we should not expect the laws to be rendered laws by their natural necessity.

That all necessity or possibility (except perhaps for the narrowly logical) is nothing more than relative necessity or possibility might be suggested by the way that modals function in ordinary language. Lewis writes:

> To say that something can happen means that its happening is compossible with certain facts. *Which* facts? That is determined, but sometimes not determined well enough, by context. An ape can't speak a human language—say, Finnish—but I can. Facts about the anatomy and operation of the ape's larynx and nervous system are not compossible with his speaking Finnish. The corresponding facts about my larynx and nervous system are compossible with my speaking Finnish. But don't take me along to Helsinki as your interpreter: I can't speak Finnish. My speaking Finnish is compossible with the facts considered so far, but not with further facts about my lack of training. What I can do, relative to one set of facts, I cannot do, relative to another, more inclusive, set.[33]

On this picture, any modality is relative to some contextually determined, typically tacit "conversational background" B. The simplest version of this idea is that B picks out some of the facts and p possesses

B-necessity exactly when *p* follows logically from those facts.[34] This approach allows us to make sense of many ordinary modal remarks that do not employ one of the philosophically venerable modalities (such as logical, mathematical, metaphysical, and natural modality). For example (suggesting the remark's context parenthetically):

- Jockl must sneeze (in view of the present state of his nose and so forth).

- My car can go from 0 to 60 miles per hour in 3.9 seconds.

- Pavarotti can't sing tonight. (He has a sore throat.)

- These two individuals, despite being opposite-sex members of the same biological species, cannot interbreed (considering their places in the group's social-dominance hierarchy).

- I cannot keep our appointment tonight.

- This box cannot hold more than 6 ounces of prunes (even if you pitted, mashed, and compressed them).[35]

Lewis's Best System (or Armstrong's sui generis relation's holding among universals) *can* supply certain facts with a variety of necessity *if* any variety of necessity is merely relative necessity. For then there is such a thing as necessity relative to the facts in the Best System (or about universals).

But what, on this view, should we make of the philosophically venerable necessities? If every variety of necessity is nothing more than membership in some set of truths that has been invoked (generally tacitly), then the necessities possessed respectively by the natural laws, the mathematical truths, and the logical truths are no more substantial than some variety of necessity possessed by a given arbitrary fact, such as (in the above example) that this box does not hold 7 ounces of prunes. This picture of necessity supplies no reason for natural (or any other kind) of necessity to be more significant than necessity relative to some arbitrary class of facts. Yet we do privilege the philosophically venerable varieties of necessity. We expect them to perform metaphysical feats that we do not demand of merely conversational necessities.

For example, suppose it is a fundamental law of nature (Coulomb's law) that between any two point bodies long at rest, *R* centimeters apart, and electrically charged to *Q* and *Q'* statcoulombs (respectively), there is a mutual electrostatic repulsion of $Q Q'/ R^2$ dynes, for any values of

Q, Q', and R. That this is a law explains why there is such a force between the bodies in every such pair, throughout the universe's history. It is no accident or coincidence that in every such case, there is such a force. This regularity does not reflect some special conditions that just happened to prevail on every such occasion. Even if charge pairs had existed under different conditions, this regularity p would still have held. The reason p holds is that p is required by law. If Coulomb's law is fundamental, then the law requiring p is none other than that p is a law.

By entailing that p was unavoidable (that is, that p would still have obtained, under every naturally possible circumstance), p's natural necessity explains why p obtains. Therefore, p's natural necessity must be ontologically able to support p's weight. Natural necessity must then be a substantial, genuine kind of necessity. Analogous remarks apply to other flavors of genuine necessity. For example (as I mentioned in chapter 1), the fact that 23 cannot be divided evenly by 3 explains why each time mother tries to divide 23 strawberries equally among her three children without cutting any (strawberries), she fails.

In contrast, suppose that Pavarotti fails to sing at the gala at which he was scheduled to perform. In a given context (say, to the audience waiting in the theater), an explanation of his not singing might be that he could not sing, where this modality is understood to be relative to certain sorts of facts about his current physical condition. (These facts include that his throat is sore; the explanation informs us that his failure to sing was not the result of a temper tantrum or a missed airline connection.) However, in some other context, this modality is not salient; these facts about his throat's current state possess no salient kind of necessity and cannot explain his failure to sing. For instance, hearing about Pavarotti's failure to sing that evening, Pavarotti's physician would seek an explanation that also *explains* the current state of his throat rather than taking that state as necessary. What suffices in the audience's context to make Pavarotti's failure to sing "unavoidable" does not make it "unavoidable" in the physician's context. In contrast, even in a context where a given law's natural necessity fails to explain why it holds (because it is a derivative law, let's say, and the context demands an explanation deriving it from certain deeper sorts of laws), the explainer's natural necessity still renders what's explained unavoidable. The explanatory significance of a merely conversational modality is just as temporary as the context that brings it into salience. Mathematical necessity and natural necessity, in contrast, are explanatorily potent in any context (even where a merely conversational necessity also suffices for explanatory power).

We recognize a short list of the restricted modalities as genuine—an elitism that is manifested in the metaphysical weight that we ascribe to them alone (in treating them as having a certain explanatory significance, for example).[36] But if all necessities are merely relative to some or another set of conversationally relevant facts, then it seems arbitrary for us to regard p's natural necessity as really making p inevitable but q's "necessary" relative to some other class of facts as failing to make q inevitable. Natural necessity and the other philosophically venerable varieties deserve their privileges only if they have something in common that sets them apart from all merely relative necessities.

Why is natural necessity a genuine variety of necessity? Not merely because it involves being a logical consequence of the laws. Unless the laws have already somehow earned their genuine necessity, nothing can inherit genuine necessity from them.

Similar issues arise in connection with other philosophically venerable varieties of necessity. For example, a Tarski-style analysis of logical necessity might say that a logical necessity is a truth that "remains true under all reinterpretations of its components other than the logical particles"[37] or that a logical necessity is "a sentence for which we get only truths when we substitute sentences for its simple sentences [that is, for its sentences containing no logical vocabulary]."[38] These analyses presuppose a distinction between logical and nonlogical vocabulary, for which Tarski offers no ground.[39] In that event, it seems like *any* selection of vocabulary to privilege (that is, to hold fixed under reinterpretation or intersubstitution) should yield a corresponding variety of necessity. Why, then, does *logical* necessity carry special weight? For there to be something important about logical necessity, there would have to be some independent reason why logical vocabulary carries special significance.

If some but not all of the relative necessities are genuine necessities, then necessity relative to the Best System may fail to constitute genuine necessity. In the next few sections, I will identify a feature of some modals in natural language—a feature that we pretheoretically recognize as characterizing genuine modality and that distinguishes the philosophically venerable modalities from the merely conversational ones.[40] Thus we will see why the philosophically venerable necessities (such as natural necessity) are varieties of genuine necessity. With natural necessity as *something* independent of lawhood, p's natural necessity is available to make p a law—in answer to the Euthyphro question.

2.7. The Modality Principle

Robert Stalnaker and Tim Williamson have independently suggested that p is necessary exactly when $\sim p \;\square\!\!\rightarrow\; p$ (or, equivalently, exactly when $q \;\square\!\!\rightarrow\; p$ for any q).[41] This proposal presumes that counterfactuals with impossible antecedents are (vacuously) true. This presumption is commonly enshrined in formal logics of counterfactuals. (If there are *no* possible q-worlds, then trivially, p holds in *every* possible q-world.) However, counterfactuals of the form "If p weren't the case, then p would still be the case" are not in ordinary or scientific use. Consequently, we should be reluctant to presume them all true.

Indeed, contrary to Stalnaker and Williamson, some counterfactuals with impossible antecedents are not trivial, but function much like other counterfactuals in natural language. For example, the counterlogical "Had there been a violation of the principle of double negation, then Gödel would probably have discovered it" is true in certain contexts where "... then I would probably have discovered it" is false, and neither is trivially so. The same applies to countermathematicals. Had Fermat's last theorem been false, then a computer program searching for exceptions to it might well have discovered one, but it is not true that I would have done so. Augustine was not aiming to make a trivial point in asserting that had six not been a perfect number, then God's acts of creation would not have occupied six days.[42] Thus, Stalnaker's and Williamson's suggestion fails to identify a natural-language marker of genuine necessity.

Furthermore, even if Stalnaker and Williamson were correct that counterpossibles are trivially true, the biconditional "p is necessary exactly when $\sim p \;\square\!\!\rightarrow\; p$" would not generalize to any variety of necessity other than the strongest kind. For example, Stalnaker's and Williamson's proposal cannot account for the natural laws' possessing a variety of necessity that is weaker than logical, mathematical, conceptual, and metaphysical necessity, but nevertheless genuine. If p is

Coulomb's law, then it is not the case that $\sim p \,\square\!\!\rightarrow p$. Our aim is to discover how genuine modalities differ from merely relative modalities without collapsing all of the species of genuine modality into one.

However, I think that Stalnaker and Williamson point us in the right direction. To distinguish genuine necessities from merely relative ones, let's look for differences in their behavior in counterfactuals. To begin with, consider the following principle:

> Whatever would have happened, had something possible happened, is also possible.[43]

In other words, for any p and q:

> If $\lozenge p$ and $p \,\square\!\!\rightarrow q$, then $\lozenge q$.

This principle seems intuitive and enshrined in ordinary reasoning. For instance, if it is possible for a large number of guests to arrive for lunch today, and we would run out of food if a large number of guests were to arrive for lunch today, then it is possible that we will run out of food. Likewise, all of the oxygen molecules in the room I now occupy could (as a matter of natural possibility) move to the far side of the room and remain there for the next few minutes. Were this exceedingly unlikely event to occur, I would lose consciousness despite being in a room with plenty of air. So it is naturally possible for that to happen.

These two examples involve different species of modality. I suggest that the above principle holds for every species of modality (see note 5). The "$\lozenge$" in a given instance of the principle can refer to any species of modality, as long as every token "$\lozenge$" in that instance refers to the same species. Although the truth-values of counterfactuals are notoriously context-sensitive, the above principle is meant to hold in any context. It says (for any p and q) that in any context where $\lozenge p$ and $p \,\square\!\!\rightarrow q$, it is also the case that $\lozenge q$.[44] The counterfactual $p \,\square\!\!\rightarrow q$ is being considered in the same context where the salient variety of modality deems p to be possible. (I'll say more about this a bit later.)

We implicitly invoke this principle when we use $\lozenge p$ and $p \,\square\!\!\rightarrow q$ to demonstrate that $\lozenge q$. For example, how might we argue that the British attack in the Battle of the Somme could have succeeded (though it failed)? We might argue that had each British soldier carried a lighter pack (each staggered under 66 pounds of equipment), then the British would have moved more rapidly across "no man's land," and so the Germans would have had no time to man their machine guns before the British overwhelmed them.[45] This argument may show that British success was possible (in terms of the conversationally salient modality), but only because it was possible (in terms of that modality) for each Tommy to have been issued a lighter pack that day. However, we cannot show that British success was possible by arguing that had the British soldiers possessed the fleetness of gazelles, then the Germans would have had no time to man their machine guns before being overwhelmed—for even if that counterfactual conditional is true, its antecedent is not possible (in terms of the conversationally salient modality), and so the conditional's truth fails to demonstrate its conclusion to be possible.

The above principle follows immediately from a slightly stronger but (I think) also plausible principle:

In any context: whatever might have happened, had something possible happened, is also possible.

In other words:

In any context: if $\Diamond p$ and $p \; \Diamond\!\!\rightarrow q$, then $\Diamond q$.

For example, if it is possible that Uncle Phil will show up for lunch today, and we might hear one of his fishing stories if he were to show up, then it is possible that we will hear one of his fishing stories. Likewise, that the Germans might have been too late in manning their machine guns, had the British carried lighter packs, is enough to show that British success was possible (since it was possible for the Tommies to have been issued lighter packs).

The above principle is equivalent to

In any context: if $\Diamond p$ and $\sim\!\Diamond q$, then $\sim [p \; \Diamond\!\!\rightarrow q]$.

Now $\sim\!\Diamond q$ is equivalent to $\Box\!\sim\!q$, so

In any context: if $\Diamond p$ and $\Box\!\sim\!q$, then $\sim [p \; \Diamond\!\!\rightarrow q]$,

or, more perspicuously, what I shall call the "modality principle" (M):

M   In any context: if $\Diamond p$ and $\Box q$, then $\sim [p \; \Diamond\!\!\rightarrow \sim\!q]$.

M concerns, for any context, whatever relative modality is salient there—that is, necessity (and possibility) relative to whatever class of facts B is implicitly designated there.

M says that if $\Box q$, then $\sim [p \diamond\!\!\rightarrow \sim q]$ for any p where $\diamond p$. Of course, if it is *not* necessary that q, then $\sim q$ is possible, and so $[p \diamond\!\!\rightarrow \sim q]$ holds for some p where $\diamond p$—namely, where $\sim q$ serves as p. Hence, M holds in both directions:

In any context: $\Box q$ if and only if $\sim (p \diamond\!\!\rightarrow \sim q)$ for any p where $\diamond p$.

We might interpret $\diamond p$ so that the above is equivalent to

In any context: q is necessary if and only if $\sim (p \diamond\!\!\rightarrow \sim q)$ for any p that is logically consistent with all of the n's (taken together) where it is necessary that n.

This complicates matters slightly, since logical necessity is now involved along with whatever flavor of necessity was originally under discussion. However, the above result is usefully compared to NP from chapter 1 (as extended to demand not only that the laws would still have held, but also that none of their negations might have held, under the relevant range of counterfactual suppositions):

m is a law if and only if in any context, $\sim (p \diamond\!\!\rightarrow \sim m)$ for any p that is logically consistent with all of the n's (taken together) where it is a law that n.

Roughly speaking, M's holding in both directions generalizes NP to all flavors of necessity. More precisely, if the laws are exactly the natural necessities, and natural necessity must qualify as a variety of necessity in every context, then NP follows from M's holding in both directions.

M demands that a necessity of a given variety be preserved under any counterfactual (or subjunctive) supposition p expressing a possibility of that variety. Thus M contrasts with Stalnaker's and Williamson's proposal, which unrealistically demands that a necessity be preserved under any counterfactual supposition whatever.

That being necessary involves being unavoidable, inescapable, or inevitable has often been elaborated in terms approximating M. For example, John Stuart Mill writes:

If there be any meaning which confessedly belongs to the term necessity, it is *unconditionalness*. That which is necessary, that which *must* be, means that which will be whatever supposition we make with regard to other things.[46]

A necessary truth is unconditional, like (my daughter, Rebecca, has suggested) unconditional love: there are no possible circumstances under which it would be lost. The necessities would all still have held under any "supposition we make with regard to *other things*," that is, with regard to things other than the necessities—which I interpret as any supposition that does not contradict the necessities.

M makes good sense under a possible-worlds construal of counterfactual conditionals. If we understand $\Diamond p$ to be true exactly when there is a possible p-world (for the given species of possibility) and $\Box q$ to be true exactly when q holds in every possible world (for the same species of possibility), then M says that relative to the actual world, any *possible* world is closer than every *impossible* world (for a given species of possibility). This notion seems implicit in the very idea of possibility and impossibility.[47] It would be strange indeed for some impossible world (for a given species of possibility) to qualify as closer than—or even equally close as—some possible world. (What, then, would its impossibility amount to?)

2.8. A Proposal for Distinguishing Genuine from Merely Relative Modalities

We were considering the argument that "necessity" relative to (say) Lewis's Best System qualifies as genuine necessity because any variety of genuine necessity is nothing more than necessity relative to some or another designated class of facts. On this view, a truth's natural necessity cannot be responsible for its lawhood since natural necessity is just necessity relative to the laws. I reject this view; I contend that natural necessity and other species of genuine necessity differ from merely relative necessities. In particular, I shall now suggest that M can be used to distinguish genuine from merely conversational necessities. In this way, M reveals why natural necessity and the other philosophically venerable varieties of necessity are all species of the same genus. A truth's natural necessity can then be responsible for making it a law.

Let's see how M distinguishes genuine from merely relative modalities. Take one of my earlier examples: This box cannot hold more than 6 ounces of prunes. We can easily imagine a conversational context in which the fact n that this box does not hold more than 6 ounces of prunes qualifies as necessary in terms of the salient relative modality. For example, suppose that you (a prune wholesaler) are discussing how to ship prunes to me (a prune retailer). You might say (alluding to the box under discussion—the kind that you use to ship prunes), "The box cannot hold more than 6 ounces of prunes." I might ask, "What if they were pitted or mashed?" You could reply, "Even if they were pitted or mashed, the box would still not hold more than 6 ounces of prunes; it is not the case that the box might have held more than 6 ounces, had the contents been pitted or mashed." This result accords with M (where p is that the prunes are pitted or mashed): If $\Diamond p$ and $\Box n$, then $\sim [p \Diamond\!\!\rightarrow \sim n]$.

There are many similar counterfactual suppositions, such as "Had the prune packers worn orange shirts and said a little prayer before packing the prunes." None of that would have made any difference to the weight of prunes that could be packed into the box. But consider "Had each prune been the size of a pea," or "Had each prune been 1 cubic millimeter in volume but the same weight," or "Had prunes possessed the density of lead" (or "Had prunes been made of lead"). In the given context, the wholesaler might reply, "If life were only so simple! Of course, if prunes were made of lead, then we would have no trouble packing more than 6 ounces of prunes into the box." Here we have various counterfactual suppositions p where it is not the case that $p \Box\!\!\rightarrow n$, and so it is the case that $p \Diamond\!\!\rightarrow \sim n$. (I discussed this relation between might-conditionals and would-conditionals in chapter 1.) But M says that if $\Diamond p$ and $\Box n$, then $\sim [p \Diamond\!\!\rightarrow \sim n]$. So for M to be satisfied, given $\Box n$, each of these p's must fall outside of M's range because $\sim\Diamond p$, that is, $\Box \sim p$. In other words, the necessities (of the salient variety) must include various facts about the density and chemical composition of prunes, not to mention various natural laws, conceptual and mathematical truths, and logical truths. (After all, had the laws of nature been different, then more than 6 ounces of prunes might have fit into the box.)

Other facts are likewise necessary. Had the horticultural history of the plum been different, then the density of prunes today might have been different (it is about 0.72 grams per milliliter), and so the box might have held more than 6 ounces of prunes. For that matter, had the box been larger, then it might have held more. Had the machine that manufactured the box been different, then the box might have been larger and so held more. To uphold M, even more facts must possess the salient variety of necessity: facts about the plum's horticultural history, the box's volume, and the box-manufacturing machinery.

Indeed, it is difficult to think of a fact that is relevant in the given conversational context but can afford to lack the salient variety of necessity without violating M. Suppose that you are selling prunes wholesale at about $1 per box. Had the price per box been different, would the box still have held 6 ounces of prunes or less? Perhaps not: had the price per box been different, the price might have been higher because the box contained more than 6 ounces of prunes. If $(p \;\Box\!\!\to n)$ is false, where p is that the wholesale price of a box of prunes differs from about $1, then M demands that $\sim\!\Diamond p$ and hence that $\Box\!\sim\!p$. Thus the fact that your price per box is $1 must qualify as necessary, on pain of violating M.

Admittedly, for this p, there are circumstances and contexts where n qualifies as necessary in terms of the salient relative modality and yet $(p \;\Box\!\!\to n)$ is true—where, for example, had the price per box differed from about $1, then that price difference would have resulted exclusively from your having had higher production costs per ounce of prunes, not because the box contained more than 6 ounces of prunes. But even in such a case, there are still plenty of routes by which M may ultimately demand that $\Box\!\sim\!p$. For instance, we just saw that to uphold M, certain facts about the prune's density must possess the salient variety of necessity; had the density of prunes not been about 0.72 grams per milliliter, then the box might have held more than 6 ounces of prunes. Had the price per box differed from about $1 because of the higher production costs per ounce of prunes, those higher costs might have resulted from a diminished prune supply as a result of global warming, and the higher temperatures and diminished rainfall in plum-growing regions might also have resulted in plums with a greater sugar content, and hence in prunes with a density considerably less than 0.72 grams per milliliter. Once again, if a necessity (a fact about the density of prunes) fails to be preserved under the supposition that the wholesale price per box differs from about $1, then M requires that the $1 wholesale price be necessary.

Notice how readily M promotes the spread of the salient variety of necessity among the relevant facts. Given that one fact n (that the box contains no more than 6 ounces of prunes) possesses that variety of necessity, other facts m (such as that the prune's density is about 0.72 grams per milliliter) must also be necessary, since it is not the case that $(\sim\!m \;\Box\!\!\to n)$ in this context. Then given that m is necessary, still other facts p (such as that the price per box is about \$1) have to be necessary, since it is not the case that ($\sim p$ $\square$→ m) in this context. To give another example: had the prunes in the box been mashed, then although (we said) the box would still have contained no more than 6 ounces of prunes (n), the price per box might have differed from about \$1 to cover the extra expense of mashing. In that event, necessity ripples outward to cover the fact that the prunes are not mashed—when M is applied not to the necessity of $n,$ but rather to the necessity of p (that the wholesale price per box is about \$1).

I conjecture that this chain reaction ramifies indefinitely, rippling outward to cover *all* of the relevant facts. That is, I propose that by this sort of reasoning, *every* fact that is relevant to the conversation ends up qualifying as necessary in terms of the species of modality that is salient in the conversation. The relative modality fails to divide the relevant facts into necessities and contingencies. Rather, to comply with M, all of the relevant facts must be necessities. In other words, when a merely relative modality divides the facts into the necessities and the contingencies, either M fails or the division turns out to be between the facts that we care about in the given context and those we do not.[48] For example, in the context of our prune-sale conversation, the handedness, hair, and height of the person who packed the prunes I ordered are matters we do not care about, and compliance with M does not ramify far enough to make them necessary: any truth possessing the variety of necessity salient in our prune-sale conversation would still have held, even if the prunes I ordered had been packed by a left-handed, red-haired six-footer.[49]

My conclusion might seem too extreme. Surely, you may say, when a potential prune customer thinks about how many boxes to buy, she treats certain relevant facts (such as the size and price of the boxes) as fixed constraints within which she must make her decision—that is, as necessary—and she treats other relevant facts (such as the number of boxes she will eventually buy) as open, and hence contingent (in terms of the salient modality). Thus, not all of the relevant facts count as necessary. However, I reply, suppose the customer would have purchased 200 boxes only if the constraints (such as the price) had been different. So for the constraints, as necessities, to comply with M, the fact that the customer does not purchase 200 boxes must also count as necessary. (Of course, the customer may not realize that this fact is necessary before she completes her deliberations.)

In contrast to a merely relative modality, a genuine modality upholds M without every relevant fact having to be necessary. The box of prunes I am ordering cannot (in view of the natural laws) be shipped faster than the speed of light. Had the price been different or the weather in plum-growing regions been different, then the laws would still have held. So M exerts no pressure to expand the range of natural necessities to include facts about the price or the weather.

2.9. Borrowing a Strategy from Chapter 1

I have conjectured that for a merely relative modality to comply with M, the division between necessities and contingencies must be the division between relevant and irrelevant facts in a conversational context where that modality is salient. This conjecture gains support from an argument reminiscent of chapter 1.

Suppose for the sake of *reductio* that some fact p is relevant in the given conversation but not necessary in terms of the salient modality. Suppose that some other fact q is relevant and necessary, where p neither logically entails nor is logically entailed by q. Now consider the counterfactual supposition ($\sim p$ or $\sim q$) entertained in the given context. By M, ($\sim p$ or $\sim q$) $\Box\rightarrow q$, and so ($\sim p$ or $\sim q$) $\Box\rightarrow \sim p$. This seems intuitive—if the modality is on the short list of genuine modalities. For instance, had 23 been divisible evenly by 3 or mother divided the strawberries evenly (without cutting any) among her children, then 23 would still have been indivisible evenly by 3, and so mother would have to have had either a different number of children than 3 or a different number of strawberries than 23.

But suppose instead that the salient species of modality is merely relative. In picturesque terms, the supposition ($\sim p$ or $\sim q$) pits p against q, as far as remaining true is concerned. They cannot both be preserved under this counterfactual supposition; at least one must go. With p and q both relevant in the given context (and neither logically stronger than the other), the salient criterion of closeness cannot make a definite choice between them; neither takes priority over the other. Hence, p holds in some of the optimally close ($\sim p$ or $\sim q$)-worlds, but q holds in others. Therefore, in the given context, ($\sim p$ or $\sim q$) $\Diamond\rightarrow p$ and ($\sim p$ or $\sim q$) $\Diamond\rightarrow q$ hold. But the former entails ($\sim p$ or $\sim q$) $\Diamond\rightarrow \sim q$, contrary to M's demand that $\sim [(\sim p$ or $\sim q)$ $\Diamond\!\!\rightarrow \sim q]$. *Reductio* achieved. Thus, to comply with M, every fact that is relevant in the given conversation must be necessary in terms of the salient variety of modality.

Admittedly, the arguments I have just given for my conjecture are hardly decisive. That both p and q are relevant in the given context does not show that the salient criterion of closeness cannot make a definite choice between them; perhaps p is more important to closeness than q. That under pressure from M, many more facts than we might have expected turn out to qualify as necessary (in terms of the modality in play during our prune-sale conversation) does not show that *all* relevant facts are necessary. For that matter, the meaning of "all relevant facts" remains rather vague. Furthermore, several species of relative modality can be in play over the course of the same conversation. If each is *merely* relative rather than a genuine species of modality, then according to my conjecture, each deems necessary exactly the relevant facts. But these are exactly the same facts, unless the range of relevant facts shifts over the course of the conversation—so my conjecture entails that the supposedly multiple species of merely relative modality are not distinct after all. Let's look briefly at these objections.

Perhaps the range of relevant facts does indeed shift in perfect time with shifts in the salience of various relative modalities over the course of a conversation. Imagine the following meeting in the White House:

PRESIDENT	What can we do about the threat of avian flu? Is it possible for us to manufacture 100 million vaccines by next year?
SECRETARY OF HEALTH	It is possible for us to manufacture the vaccine in such large quantities. We have the scientific and engineering know-how.
PRESIDENT	So ordered!
SECRETARY OF HEALTH	Um, Mr. President, I'm afraid it's quite impossible for us to carry out your order; we don't have the requisite production capacity or trained personnel or raw materials.

PRESIDENT What about if our level of commitment were on the scale of the Manhattan Project?...

Plausibly, the Secretary of Health regarded our production capacity as relevant by the time she made her second remark, once the President had issued the order—whereas the Secretary initially regarded only our level of technological know-how as relevant. Perhaps mention of the Manhattan Project was about to shift the range of relevant facts again.[50]

Another complication that I have ignored so far is that talk of "the conversational context" fails to distinguish two contributions that the context makes: to the truth-values of counterfactuals and to the salience of given species of modality. Context might shift enough to change a counterfactual's truth-value without changing which relative modality is salient. Let us return to the prune-sale conversation. In the context where we agree that the box (sitting prune-filled at our feet) cannot hold more than 6 ounces of prunes, I may say, "Had your price per box been much higher than \$1, then I might not have bought the box, but then again, I might still have bought it because the price might have been higher because the box was filled with more than 6 ounces of prunes." However, somewhat later in our conversation, the context may shift so that "Had your price per box been much higher than \$1, a box would still have contained no more than 6 ounces of prunes and so I would not have bought the box" is true.

That context can shift so as to change the truth-values of counterfactuals, while a given relative modality remains salient, allows me to strengthen my *reductio* argument by weakening one of its premises. The argument I gave earlier presumed that in *any* context where both p and q are relevant (and p neither logically entails nor is logically entailed by q) and $\Box q$ and $\sim\Box p$ hold in terms of the salient merely relative modality, ($\sim p$ or $\sim q$) $\Diamond\!\!\to \sim q$ holds, contrary to M's demands. That is, the salient criterion of possible-world closeness cannot make a definite choice between the importance of preserving p and the importance of preserving q; neither takes priority over the other. Earlier, I wondered whether this was correct. However, having now distinguished the context's contribution to making counterfactuals true from the context's contribution to making some relative modality salient, I can clarify M: it demands that if the context makes salient a given species of modality in terms of which $\Diamond p$ and

$\square q$ hold, then $\sim (p \diamond \!\!\!\rightarrow \sim q)$ holds no matter how the context may influence the truth-values of counterfactuals. So rather than having to presume that $(\sim p$ or $\sim q)$ $\diamond \!\!\!\rightarrow \sim q$ holds in *all* of the contexts where the given relative modality is salient, my argument requires merely that in *some* such context, $(\sim p$ or $\sim q)$ $\diamond \!\!\!\rightarrow \sim q$ holds. That is enough to run afoul of M.

In other words, M requires that q's preservation take priority over p's preservation in *every* context where the given variety of modality is salient. This is plausible for the short list of genuine modalities, but not for the others. Thus genuine modalities function differently from merely relative ones.

2.10. Necessity as Maximal Invariance

I have just argued that for a merely relative modality, the necessities encompass every fact that is relevant in a conversational context where that species of modality is salient (on pain of violating M). That argument was essentially the same as chapter 1's argument that a set of sub-nomic truths containing an accident cannot attain sub-nomic stability without including every sub-nomic truth. The similarity between these arguments is not coincidental. It arises from the fact that according to M (once it is supplemented with nested counterfactuals), the sub-nomic truths possessing whatever kind of relative necessity is salient in a given context must form a set that behaves in that context exactly as a sub-nomically stable set would.

We saw earlier that M might be interpreted as

> In any context: q is necessary if and only if $\sim (p \diamond \!\!\!\rightarrow \sim q)$ for any p that is logically consistent with all of the n's (taken together) where it is necessary that n.

So for a nonempty set Γ of sub-nomic truths containing every sub-nomic logical consequence of its members, M entails that in a given context, Γ contains exactly the necessities only if for each member m of Γ, $\sim (p \diamond \!\!\!\rightarrow \sim m)$ holds for any sub-nomic claim p where $\Gamma \cup \{p\}$ is logically consistent. Now suppose that a given variety of necessity is genuine (rather than merely relative to some conversationally salient set of facts) if and only if it satisfies M in every context. Then if we include the nested counterfactuals,[51]

Γ contains exactly the necessities, for some genuine variety of necessity, only if for each member m of Γ (and in every conversational context),

$\sim (p \lozenge \to \sim m)$,

$\sim (q \lozenge \to (p \lozenge \to \sim m))$,

$\sim (r \lozenge \to (q \lozenge \to (p \lozenge \to \sim m)))$,...

for any sub-nomic claims $p, q, r, \ldots$ where $\Gamma \cup \{p\}$ is logically consistent, $\Gamma \cup \{q\}$ is logically consistent, $\Gamma \cup \{r\}$ is logically consistent, and so forth.

That is, Γ contains exactly the necessities, for some genuine variety of necessity, only if Γ possesses "sub-nomic stability" as defined in chapter 1.

We have, at last, arrived at this chapter's main thesis: for each variety of genuine necessity, the sub-nomic truths possessing it form a sub-nomically stable set—and for each sub-nomically stable set (except the set of all sub-nomic truths, if it be stable), there is a variety of genuine necessity where the sub-nomic truths so necessary are exactly the set's members. In short, for the sub-nomic truths, there is a correspondence between the varieties of genuine necessity and the nonmaximal sets possessing sub-nomic stability.[52]

Intuitively, "necessity" is an especially strong sort of persistence under counterfactual perturbations. But not every fact that would still have held, under even a wide range of counterfactual perturbations, qualifies as possessing some species of "necessity." No necessity of any genuine kind is possessed by an accident, even an accident that would still have held under many counterfactual suppositions. Being "necessary" is supposed to be *qualitatively* different from merely being invariant under a wide range of counterfactuals. Necessity involves being inevitable, unavoidable—being the case no matter what, in the broadest possible sense of "no matter what."

We can capture this idea by identifying the varieties of necessity among the sub-nomic truths with the nonmaximal sub-nomically stable sets. By definition, a set is sub-nomically stable if and only if its members are together invariant under as broad a range of sub-nomic counterfactual suppositions as they could together be. Trivially, they could not all be preserved under a supposition with which one of them is logically inconsistent, but their sub-nomic stability requires their preservation under every other sub-nomic supposition. So a sub-nomically stable set is *maximally* resilient—as resilient as *it* could logically possibly be (as far as sub-nomic suppositions are concerned). Accordingly, its members possess a variety of necessity. No species of necessity is possessed by an accident, even by one that would still have held under many counterfactual suppositions, since it does not belong to a nonmaximal set of sub-nomic truths having as much invariance under sub-nomic counterfactual suppositions as it could logically possibly have.[53]

The identification of necessity with stability can be motivated by the principle that began section 2.7:

> Whatever would have (or, more broadly, might have) happened, had something possible happened, is also possible.

According to this principle, the actual world is nearer to every p-world that is *possible* than to any p-world that is *impossible* (in terms of the given variety of modality). As I mentioned earlier, it would make a mockery of "possibility" for some p-world to be possible, but for the closest p-world (or even one of the optimally close p-worlds) to be impossible.

Now suppose that the necessities of some particular variety (such as logical, mathematical, moral, or natural) are exactly the members of some particular logically closed set of truths. What must that set be like? The principle from section 2.7 says that if p is possible (that is, logically consistent with the relevant set) and if m would have held, had p been the case, then m must be possible (that is, logically consistent with that set). This is immediately guaranteed if the set is stable. If p is logically consistent with the given stable set, then under the counterfactual supposition that p holds, the set's members would still have held (since the set is stable), and so anything else that would *also* have been the case must join the set's members and so must be logically consistent with them.

On the other hand, look what happens if a logically closed but *unstable* set of truths contains exactly the necessities of some variety. Because the set is unstable, there is a supposition p that is logically consistent with the set but under which some member m of the set would not still have held.[54] That is to say, m's negation might have held. But m, being a member of the set, is supposed to be necessary, so it is impossible for m's negation to be true. Therefore, if an unstable set contains exactly the necessities (of some variety), then had a certain possibility (of that variety) come to pass, something impossible might have happened. This conflicts with the principle from section 2.7. In short, if an unstable set contains exactly the necessities (of some variety), then for some p, there is a *possible* world where p obtains, but the closest p-world (or, at least, one of the optimally close p-worlds) is *impossible*—which conflicts with the principle that the actual world is closer to every possible p-world than to any impossible p-world. Hence, if a logically closed set of truths contains exactly the necessities (of some variety), then the set must be stable.

The identification of necessity with stability allows us to understand how the laws manage to qualify as necessary despite being contingent. They are necessary by virtue of forming a sub-nomically stable set. Their necessity is no small accomplishment, since it is not the case that *every* cockamamie logically closed set of truths is stable. What makes the set of broadly logical truths and the set of natural laws *alike* is that they both form stable sets. It is this commonality that makes both classes of truths "necessary."

Yet they remain distinct species of necessity. The laws of nature are not as necessary as the broadly logical truths since they rank lower in the pyramidal hierarchy of sub-nomically stable sets (depicted in chapter 1, section 1.9). That is, the range of counterfactual suppositions under which the laws must all be preserved, for the set of laws to qualify as stable, is narrower than the range of counterfactual suppositions under which the broadly logical truths must all be preserved, for the set of broadly logical truths to qualify as stable. (While asserting the countermathematical that had six not been a perfect number, then God's acts of creation would not have occupied six days, Augustine also said that six would still have been a perfect number even if the works of God's creation had not existed.) Natural necessity lies between broadly logical necessity, on the one hand, and no necessity at all, on the other hand. My proposal thus gives different varieties of necessity a unified treatment, identifying something common to each in virtue of which they all qualify as species of the same genus—but without collapsing every variety of necessity into one.

An account that identifies necessity with stability can do it all: it can account for the laws' necessity *and* their contingency.

I have suggested that the broadly logical truths (in other words, the narrowly logical, conceptual, mathematical, and metaphysical truths along with the moral laws and so forth) form a sub-nomically stable set. Perhaps some proper subsets of the broadly logical truths also form sub-nomically stable sets. These sets then should appear somewhere above the set of broadly logical truths in the pyramidal hierarchy depicted in figure 1.3. For instance, perhaps narrowly logical impossibility entails metaphysical impossibility, but not vice versa, so that the narrowly logical necessities are a stable proper subset of the metaphysical necessities, which also form a stable set. Perhaps the moral necessities form a stable set that includes the conceptual and mathematical necessities. Accordingly, perhaps the narrowly logical necessities form the smallest sub-nomically stable set; they would still have been true even if the mathematical truths, for example, had been different.[55] (In that event, for instance, had the axiom of choice been false, then the cardinals would not have been linearly ordered, but every substitution-instance of $\sim(p \mathrel{\&} \sim p)$ would still have held.[56]) The broadly logical truths would then all possess a kind of necessity falling between narrowly logical necessity and natural necessity.

Among the broadly logical truths, the metaphysical truths presumably include a diverse lot: the "laws" of mereology, eternalism (or presentism, whichever is true), realism (or nominalism, whichever is true) about properties (or propositions or numbers), and so forth—all of the fundamental abstract ontological truths—along with (some philosophers say) "Water is H_2O" and its colleagues. Some of these broadly logical truths seem less metaphysical—that is, more like natural laws—than others. For example, consider space-time substantivalism (or relationalism, whichever is true), or the fact that nothing is both positively and negatively electrically charged.[57] These facts seem simultaneously like very exalted natural laws and relatively lowly metaphysical truths. The hierarchy of sub-nomically stable sets could capture this intermediate modal status by including, somewhere above the set of broadly logical truths, a more exclusive set containing all of the broadly logical truths except for these lowly metaphysical truths. Both sets lie above Λ in the pyramid, reflecting the fact that even these lowly metaphysical truths are modally more exalted than ordinary natural laws.

On the other hand, perhaps all of the broadly logical truths differ only in their subject matter, not in their necessity, because the smallest sub-nomically stable set includes them all. Perhaps the narrowly logical necessities (for example, "Nothing that has a color does not have a color") fail to form a sub-nomically stable set by themselves, but must be supplemented by metaphysical necessities (for example, "All red objects are colored").[58] So one must think if one finds the following counterfactual plausible (at least in some contexts): "Had there been something red but not colored, then there would have been something that has a color and does not have a color."

I shall remain agnostic about whether there are species of necessity possessed by some broadly logical truths but not by others. My concern is only with what it *would be* for such species to exist. Whether the narrowly logical truths possess a stronger variety of necessity than the metaphysical truths possess, and whether the metaphysical, mathematical, and narrowly logical truths all possess a species of necessity that moral laws lack, depends on the truth of various counterfactual conditionals. For instance, whether the narrowly logical truths possess a stronger variety of necessity than the metaphysical truths turns on whether the narrowly logical truths would all still have held even if red had not been a color.

In other words, my proposal explains what it would take for mathematical truths, conceptual truths, and so forth to be as necessary as the narrowly logical truths. Our uncertainty about whether these other truths possess the same kind of necessity as the narrowly logical truths may be matched by our uncertainty about the truth of various counterfactual conditionals that must hold (in some contexts, at least) for all broadly logical truths to be modally on a par.

The pyramidal hierarchy also explains why someone aiming to show that there is no such thing as metaphysical necessity should find it relevant to argue that in certain contexts, a putative metaphysical necessity (such as that water is H_2O) would not still have held under a counterfactual supposition that (albeit logically inconsistent with the natural laws) is logically consistent with all of the alleged metaphysical necessities ("Had H_2O been poisonous, nonliquid, opaque, and unable to extinguish fires and ZPQ been healthful, liquid, transparent, and able to douse fires"). The aim of this argument, as I understand it, is to show that the alleged metaphysical necessities (together with the logical, conceptual, mathematical, and moral truths—but without any ordinary natural laws) fail to form a sub-nomically stable set.

By the same token, for the broadly logical truths to possess some variety of necessity stronger than natural necessity depends upon the truth of various counterfactual conditionals. For a moral truth such as (let's suppose) "Capital punishment is immoral" to rise to the level of a moral law, rather than constitute a contingent fact, various subjunctive conditionals must be true, such as "Capital punishment would still be immoral even if it deterred crime."

In chapter 1, I suggested that certain proper subsets of Λ may be sub-nomically stable. For example, take the sub-nomic logical conse-quences of the fundamental dynamical law, the law of the composition of forces, and the conservation laws (perhaps with certain other elite laws, such as the coordinate transformations), along with the broadly logical sub-nomic truths. They may form a sub-nomically stable set without any of the force laws (such as Coulomb's law, the gravitational-force law, and so forth). This set's sub-nomic stability requires, for exam-ple, that component forces would still have combined to yield net forces in accordance with the parallelogram of forces even if the kinds of forces had been different—for instance, even if gravity had declined with the cube of the distance. If the varieties of necessity correspond to the sub-nomically stable sets, then the different "strata" of natural law are associ-ated with different varieties of necessity. While there is a kind of necessity that all and only Λ's members possess, there is also a stronger species of necessity not possessed by the force laws, but possessed by the conserva-tion laws, parallelogram of forces, and fundamental dynamical law. They are lifted clear of the ruck of the various special force laws. Termi-nologically, we might reserve "natural necessity" for the more inclusive necessity—associated with membership in the largest nonmaximal sub-nomically stable set—or we might prefer to say that there are several varieties of natural necessity. (In chapter 3, I shall return to the conserva-tion laws and their characteristic variety of natural necessity.)[59]

2.11. The Laws Form a System

As we saw earlier, Stalnaker and Williamson unpack p's necessity as p's preservation under every counterfactual supposition, and thus as p's individual achievement rather than as deriving from p's belonging to some larger whole. On my view, in contrast, p's necessity involves its belonging to a nonmaximal sub-nomically stable set. Hence, its necessity (of whatever species) is a collective affair—the accomplishment of an entire team of facts. The preservation of a given fact p under some range of suppositions renders p necessary only because of the other facts that join p to form a sub-nomically stable set.

Let's consider natural necessity in particular. Because p's lawhood depends on p's membership in a nonmaximal sub-nomically stable set, each natural law is bound up with the others. Each member of the set depends on the others to help specify the range of invariance that it has to possess in order to be a law. That is, each member of the set participates in delimiting the range of suppositions under which every member must be preserved in order for the set to be stable. The laws derive their lawhood collectively; their sub-nomic stability means that they are *together* as resilient under sub-nomic counterfactual suppositions as they could *together* be. They form a unified, integrated whole—a system. (In depicting lawhood as a team effort rather than an individual achievement, my account agrees with Lewis's and differs from Armstrong's and from scientific essentialism.)

That the various laws must "cover" for one another in this way has sometimes led scientists to discover a heretofore unknown law. Let's look at an example from classical physics. Suppose we have two charged point bodies—one having been at rest for a long time, the other having been in motion for a long time in a constant direction at a constant speed v. When the moving one streaks past the stationary one, then at its moment of closest approach (at a distance r), their mutual electric repulsion equals the product of their charges divided by $r^2 \sqrt{[1-(v^2/c^2)]}$. (The force is thus greater than the force—given by Coulomb's law—between point-bodies long at rest at a separation r.) Although this $\sqrt{[1-(v^2/c^2)]}$ factor crops up everywhere in Albert Einstein's special theory of relativity, Oliver Heaviside actually discovered this law in 1888 (17 years before Einstein published relativity theory) by deriving it from James Clerk Maxwell's electromagnetic-field equations of 1873. (Maxwell's electromagnetic theory was already relativistic before relativity came along; it's a bit of 20th-century physics that was discovered in the 19th century.) Several physicists seem to have regarded Heaviside's discovery as suggesting that the speed of light is a cosmic speed limit for charged bodies, since if v exceeds c, then $\sqrt{[1-(v^2/c^2)]}$ is the square root of a negative quantity, and so Heaviside's equation yields an imaginary number as the magnitude of the force. For example, G. F. C. Searle concluded that "it would seem to be impossible to make a charged body move at a greater speed than that of light."[60] Likewise G. F. FitzGerald wrote to Heaviside: "You ask 'what if the velocity be greater than that of light?' I have often asked myself that but got no satisfactory answer. The most obvious thing to ask in reply is 'Is it possible?'"[61]

How should we reconstruct their argument for this conclusion? Here is my suggestion. It seems doubtful that all charged bodies would still have accorded with Heaviside's equation had there been a charged body moving at superluminal speed. But a law must be preserved under every p that is logically consistent with Λ. Hence, unless Heaviside's equation is not a law (or applies only to subluminal speeds[62]), there must be a law prohibiting charged bodies from moving superluminally. Thanks to that law, the failure of Heaviside's equation to be preserved under the supposition of superluminal charged bodies fails to impugn that equation's lawhood.[63] Heaviside's law depends on the law prohibiting superluminal speeds to "cover" for it: to limit the range of invariance that it has to have in order to qualify as a law.

2.12. Scientific Essentialism Squashes the Pyramid

According to scientific essentialism, laws are metaphysically necessary: for example, being electrically charged essentially involves having the power to exert and to feel forces in accordance with certain particular laws, such as Coulomb's law and Heaviside's equation. Essentialism takes counterfactuals, such as "Had I worn an orange shirt, then gravity would still have declined with the square of the distance," to be grounded in essences (in this case, gravity's).

What about a counterfactual such as "Had I worn an orange shirt, then there would still have been gravity (rather than, for example, a force varying with the inverse-cube of the distance)"? That it is metaphysically necessary for gravity to be an inverse-square force does not ensure that gravity would still have been one of the universe's forces, had I worn an orange shirt. Accordingly, some essentialists have maintained that our *world's* essence determines what natural kinds of things exist. Brian Ellis, for instance, says that a world of the same natural kind as the actual world

> must also have the same basic ontology of kinds of objects, properties, and processes. It must, for example, be a physical world made up of particles and fields of the same fundamental natural kinds as those that are fundamental in this world. If electrons and protons are such fundamental natural kinds in this world, then they must also exist in every similar world.[64]

Thus, it is a metaphysical necessity that any world of the same kind as the actual world possesses gravity and lacks a similar inverse-cube force. Ellis writes:

> Could there be fundamental natural kinds of objects, properties, or processes existing in worlds similar to ours that do not exist in our world? In other words, could a world of the same natural kind as ours have a richer basic ontology? I think not. A world with an ontology otherwise like ours, which included some extra ingredients, could...not be a world of the same specific natural kind as ours....Worlds with different basic ontologies cannot be essentially the same.[65]

Therefore, since the counterfactual supposition "Had I worn an orange shirt" supposes a world of the same kind as the actual world, there would still have been gravity rather than a similar but inverse-cube force.

I don't find this move very appealing for four reasons that I shall give in ascending order of importance.

First, to associate the law that there is gravity (rather than a similar but inverse-cube force) with the *world's* essence seems like a desperate attempt to find *something* the essence of which could be responsible for this law. Even if gravity and the electron have essences, it is not obvious that "the world"—reality—does.

Second, even if we put this point aside and presume that the actual world has such an essence, the supposition "Had I worn an orange shirt" does not explicitly or even implicitly ask us to suppose a world of the same kind as the actual world. Gravity's essence might explain why the counterfactual "Had I worn an orange shirt, then gravity would still have declined with the square of the distance" obtains— since to be gravity, it must decline with the square of the distance. But neither the antecedent nor the consequent of "Had I worn an orange shirt, then gravity would still have existed" says anything about a world of the same kind as the actual world. The actual world's essence is relevant to supporting the counterfactual only if the counterfactual's antecedent is "Had I worn an orange shirt and the world been of the same kind as the actual world . . ."

I can make my point in another way. Ordinarily, when I grasp a dry, well-made, oxygenated match and say, "Had I struck this match, it would have lit," I do not implicitly mean "Had I struck this match and kept it dry." Rather, had I struck this match, it would have remained dry, and that is part of the reason why it would have lit. Likewise, when I say, "Had I worn an orange shirt," I do not implicitly mean "Had I worn an orange shirt and the world been of the same kind as the actual world." Rather, had I worn an orange shirt, the world would still have had exactly the same natural kinds, and so there would still have been gravity and not some inverse-cube force. The challenge is to explain why the match would still have been dry and why the world would still have had exactly the same natural kinds. To build these features into the respective counterfactual antecedents is to evade that challenge. It would be just as much of a cop out to say that "Had I worn an orange shirt" implicitly includes "and there still existed gravity and no alien forces."

Rather than suppose that the antecedent implicitly demands a world of the same kind as the actual world, Ellis proposes a theory of counterfactuals along these lines:

> To evaluate a conditional, on such a theory, we should have to consider what would happen, or be likely to happen, in a world of the same natural kind as ours in which the antecedent condition is satisfied, other things being as near as possible to the way they actually are. The proposition "if A were the case, then B would be the case" will be true on such a theory if and only if in any world of the same natural kind as ours in which "A" is true, in circumstances as near as possible to those that actually obtain, "B" must also be true.[66]

This takes me to the third of my four objections. For essentialism simply to stipulate that a world "of the same natural kind as ours" is the "closest" possible world where I am wearing an orange shirt is no better than for an advocate of the Best System Account to stipulate that

generalizations in the Best System exert special influence in determining the "closest" possible world where I am wearing an orange shirt.
Why do they *deserve* to exert this influence? I will say more about this
objection in chapter 4.

But for now, I wish to turn to a fourth objection. Earlier I suggested that certain proper subsets of the natural laws may be sub-nomically stable. For instance, I suggested that the set containing the
fundamental dynamical law, the law of the composition of forces, and
the conservation laws, among others—along with the broadly logical
truths, but without the various force laws—may be sub-nomically stable, in which case these laws possess a stronger kind of necessity than
force laws do. This set's sub-nomic stability requires, for example, that
its members would still have held even if there had been different kinds
of forces (for example, even if gravity had been replaced by a force that
declines with the cube of the distance). The trouble with scientific
essentialism is that by identifying natural necessity with metaphysical
necessity, it automatically assimilates all varieties of natural necessity
into one. It squashes the pyramid of sub-nomically stable sets, treating
all varieties of natural necessity as possessing the same strength (namely,
that of metaphysical necessity).

Consider, for instance, the counterfactual "Had gravity been replaced
by an inverse-cube force, then the conservation laws, fundamental dynamical law, and law of the composition of forces would still have held." (We
might just as well have considered "Had gravity been an inverse-cube
force…" or "Had masses attracted each other with inverse-cube forces
rather than inverse-square forces…") This counterfactual may well be true
(according to classical physics), and even if it is not, an account of natural
law should leave room for it to be true—for there to be multiple strata of
natural necessity. (In chapter 3, I will say more about the explanatory significance of these multiple strata.) How can Ellis's account deal with this
counterfactual? If the actual kinds of forces are fixed by the world's essence,
then a world where gravity is replaced by an inverse-cube force does not
belong to the same natural kind as the actual world. So Ellis's principle that
a counterfactual holds exactly when its consequent holds "in a world of
the same natural kind as ours in which the antecedent condition is satisfied, other things being as near as possible to the way they actually are"
cannot address the above counterfactual. An essentialist might contend
that this counterfactual antecedent invokes a world where gravity is

replaced by an inverse-cube force, but with an essence as otherwise similar to the actual world's as possible. However, as we will see in chapter 4, this "similarity" cannot be cashed out simply in terms of the world's having as many natural kinds as possible in common with the actual world.

To put all laws on a modal par with metaphysical necessities such as "Red is a color" (or, according to some philosophers, "Water is H_2O") is to put them all on a modal par with one another, which is incorrect if laws come in strata. An essentialist might reply that we can avoid flattening the pyramid by regarding all varieties of natural necessity as distinct varieties of metaphysical necessity. Some metaphysical necessities are simply more necessary than others. But then what is it about the essences that makes this so? If essentialism can explain which laws survive which counterfactual suppositions only by adding to the essences some structure that is precisely isomorphic to the pyramid of sub-nomically stable sets, then there seems to be little explanatory value to positing that structure in addition to the pyramid, just as positing a catalogue of powers (such as dormitive virtue), one for each kind of behavior, offers little explanatory payoff. The game just isn't worth the candle.

2.13. Why There Is a Natural Ordering of the Genuine Modalities

I have suggested that "natural necessity" counts as a variety of genuine necessity because it accords with M without encompassing every relevant truth. I have used M in this way to distinguish the genuine from the merely relative modalities. That a genuine modality does justice to M leads nicely to the thought that the species of genuine necessity correspond to nonmaximal sub-nomically stable sets.

Suppose that two species of modality (whether genuine or merely relative) are in play in the same conversational context, the interlocutors distinguishing them by using locutions like "possible in view of…" Let Γ entail all and only the truths possessing one of these species of necessity, and let Σ entail all and only the truths possessing the other. Just as we showed in chapter 1 that for any two sub-nomically stable sets, one must be a proper subset of the other, so likewise we can use M to show that Γ must entail or be entailed by Σ:

Suppose (for the sake of *reductio*) that Γ possesses one species of necessity ("Γ-necessity") and Σ possesses a different species of necessity ("Σ-necessity"), both of which are salient in the given context; t is logically entailed by Γ but not by Σ; and s is logically entailed by Σ but not by Γ.

Then ($\sim s$ or $\sim t$) is logically consistent with Γ, so ($\sim s$ or $\sim t$) is Γ-possible.

By M, every Γ-necessity would still have been true, had it been the case that ($\sim s$ or $\sim t$)—where this counterfactual is entertained in the given context.

In particular, then, ($\sim s$ or $\sim t$) $\Box\!\!\rightarrow$ t holds in the given context.

Since (in the given context) t and ($\sim s$ or $\sim t$) would have held, had ($\sim s$ or $\sim t$), it follows that (in the given context) $\sim s$ would have held, had ($\sim s$ or $\sim t$). That is, ($\sim s$ or $\sim t$) $\Box\!\!\rightarrow$ $\sim s$.

Now let's apply similar reasoning to Σ. Since ($\sim s$ or $\sim t$) is logically consistent with Σ, ($\sim s$ or $\sim t$) is Σ-possible, and so by M it is not the case, for any Σ-necessity, that its negation would (or even might) have held, had ($\sim s$ or $\sim t$) been the case—where this counterfactual is entertained in the given context.

In particular, then, $\sim$ [($\sim s$ or $\sim t$) $\Box\!\!\rightarrow$ $\sim s$] holds in the given context.

But we have now reached a contradiction: ($\sim s$ or $\sim t$) $\Box\!\!\rightarrow$ $\sim s$ and $\sim$ [($\sim s$ or $\sim t$) $\Box\!\!\rightarrow$ $\sim s$]. *Reductio* accomplished.[67]

This argument shows that if two modalities (whether genuine or merely relative) are simultaneously in play, then they cannot be *utterly* different; the necessities by the lights of one must be a proper subset of the necessities by the lights of the other.

This argument has a nice payoff: it explains why there is a natural ordering of the modalities on the short list of genuine modalities by appealing to the fact that for any pair of them, there is some context where both are in play. The traditional picture of the genuine modalities involves a series of concentric circles: the (narrowly) logical necessities are a proper subset of the conceptual necessities, which are a proper subset of the metaphysical necessities, which are a proper subset of the natural necessities. This natural ordering does not extend to encompass the merely relative modalities, which come in no natural ordering. Here is another difference between the genuine and merely relative modalities.

That the genuine modalities fall into a nested sequence requires an explanation. (I know of no attempt to explain it.) By the above argument, it is explained by M and the fact that any two genuine modalities can simultaneously be conversationally salient. Examples are easy to find. For instance, when we use the natural laws to predict the behavior of a physical system, we occasionally reject one of the solutions as "unphysical," saying that it is mathematically but not physically possible. For example, suppose we compute after how long a body dropped from height h (in a gravitational field imparting constant acceleration g, ignoring air resistance, the variation in gravitational acceleration with height, and so forth) will reach the ground. The relevant physical law is $h = 1/2 \ gt^2$, yielding $t = \sqrt{(2h/g)}$. Mathematically, this expression covers both the positive root and the negative root, but the negative root is unphysical, since it depicts the body as hitting the ground sometime before it was released. So as a solution, the negative root is mathematically but not physically possible.

That M accounts for the natural ordering of the traditional short list of genuine modalities lends weight to M's other implications, such as (I have argued) that any merely relative modality takes all of the relevant facts to be necessary. Of course, one might avoid that result by holding that M applies only to the genuine modalities, not to the merely relative modalities. However, in that case, I would simply offer M as the sought-after natural-language marker of the distinction between genuine and merely relative modalities.

2.14. Conclusion: Necessity, as Maximal Invariance, Involves Stability

I began this chapter by considering a Euthyphro question: Are the laws necessary by virtue of being laws, or are they laws by virtue of being necessary? It seems to me that intuitively, their necessity is what makes them laws. But that option is unavailable if natural necessity is merely a relative necessity (namely, necessity relative to the laws), since the distinction between laws and accidents must then be metaphysically prior to the distinction between what's naturally necessary and what isn't. I have offered an alternative: that each kind of genuine necessity involves maximal persistence under counterfactual suppositions, and hence amounts to membership in some nonmaximal sub-nomically stable set. Lawhood is then not needed to define natural necessity. The laws' necessity is then available to set the laws apart from the accidents, and the laws' place in the pyramidal hierarchy of sub-nomically stable sets distinguishes them from the broadly logical necessities.

I will try to reap the benefits of these proposals in the next two chapters. On the picture that I will offer, p is a law in virtue of belonging to a nonmaximal sub-nomically stable set, and the subjunctive facts (the facts expressed by subjunctive and counterfactual conditionals) making that set sub-nomically stable are ontologically primitive. Thus, p is a law in virtue of its necessity, and its necessity is constituted by its membership in a sub-nomically stable set. I have already suggested some of the payoffs of this approach—in brief, that it allows us to understand how laws manage to be genuinely necessary yet contingent. I shall uncover some other payoffs in the next chapter.

3

Three Payoffs of My Account

3.1. The Itinerary

A philosophical proposal should be fruitful. It should pay dividends beyond one's initial investment in it by having welcome consequences that were not expressly built into it, yet follow from it in a natural way. In this chapter, I display three respects in which it is fruitful to characterize the sub-nomic truths that are laws as the members of a nonmaximal, sub-nomically stable set.

Sections 3.2 and 3.3 concern the laws' immutability. After examining several unsuccessful explanations of why the laws cannot change, I argue that my account explains nicely how a temporary law (for instance, a law that expires at a given moment) would differ from an eternal but time-dependent law (a law that remains in force at all times, but treats different moments differently). Furthermore, my account identifies why temporary laws are metaphysically impossible—why the laws are immutable. In this respect, my account does better than Lewis's and Armstrong's.

Sections 3.4 and 3.5 concern meta-laws: laws that govern the laws governing sub-nomic facts. Some meta-laws (such as symmetry principles) play important scientific roles. I explain how requirements that the laws *must* obey (as a matter of meta-law) differ from regularities to which all laws governing sub-nomic facts just *happen* to conform. Then I show how my account of lawhood in terms of stability is naturally extended to cover meta-laws.

Sections 3.6 and 3.7 concern the laws' relation to objective chances. For example, if it is a law that all atoms exhibit a certain behavior, then there is never some nonzero chance of an atom's failing to do so. I argue that whereas this relation must be inserted by hand into Lewis's account, it falls naturally out of my account.

Insofar as this chapter's three topics are unrelated, the fact that my proposal pays dividends regarding each counts more strongly in its favor.

3.2. Could the Laws of Nature Change?

The natural laws are traditionally characterized as "eternal," "fixed," and "immutable."[1] Is the laws' unchanging character a metaphysical necessity? If so, then in a given possible world, there are exactly the same laws at all times (even if laws differ in different possible worlds). On the other hand, if the laws' unchanging character is *not* a consequence of what it *is* for a truth to be a law, then if there actually always have been and will be exactly the same laws, this fact is metaphysically contingent.

Occasionally, I encounter articles with provocative titles such as "Anything Can Change, Even an Immutable Law of Nature" (*New York Times*, August 15, 2001) and "Are the Laws of Nature Changing with Time?" (*Physics World*, April 2003). These articles generally concern whether certain physical parameters heretofore believed constant may really be slowly changing. Despite the sensationalistic titles, such changes need not threaten the laws' immutability. The laws at every moment may be the same—identifying the same function of time (or of some other factor) as giving the physical parameter's value at every moment.

Likewise, some cosmologists say that as the universe cooled after the Big Bang, symmetries were spontaneously broken, phase transitions took place, and discontinuous changes occurred in the values of various physical parameters (such as the strengths of certain fundamental interactions or the masses of certain kinds of particles). These changes are sometimes described as changes in the laws:

> One usually assumes that the current laws of physics did not apply [in the period immediately following the Big Bang]. They took hold only after the density of the universe dropped below the so-called Planck density, which equals 10^{94} grams per cubic centimeter.... The same theory may have different "vacuum states," corresponding to different types of symmetry breaking between fundamental interactions and, as a result, to different laws of low-energy physics.[2]

However, perhaps this "change" in the laws is better understood as involving unchanging laws such as (to give a very simple example)

(1) Between any two electrons that have been at rest, separated by r centimeters, for at least r/c seconds, there is an electrostatic

repulsion of F dynes, if the universe is no more than 10^{-10} seconds old, and f dynes $(f \neq F)$ otherwise.

Instead of citing the universe's age, the law might instead identify the critical factor as the universe's being cooler than 3×10^{15} Kelvin (K), for example.[3] Then (1)—citing the universe's age—would be an accidental truth, not a law. The electrostatic forces between electrons before and after the temperature threshold is crossed would presumably be explained by laws that do not merely specify the strengths of these particular forces. Rather, laws more fundamental than any resembling (1) would explain why 3×10^{15} K is the critical temperature for many kinds of interactions and by what process new behavior arises from the universe's crossing this threshold. If the "phase transition" is properly understood in this fashion, then it does not involve a genuine change in the laws.

However, perhaps the phase transition is properly understood differently: as involving something like

(2) Between any two electrons that have been at rest, separated by r centimeters, for at least r/c seconds, there is an electrostatic repulsion of F dynes

holding as a law before the universe is more than 10^{-10} seconds old, and

(3) Between any two electrons that have been at rest, separated by r centimeters, for at least r/c seconds, there is an electrostatic repulsion of f dynes

holding as a law thereafter $(f \neq F)$. In that case, the phase transition involves a change in the laws. Once again, I have chosen a simple example. Presumably, the laws before the universe is 10^{-10} seconds old would include (2) as a consequence of some broader law, covering more than the mutual electrostatic repulsion between stationary electrons (and likewise for the laws after the phase transition).

I will shortly examine how (1)'s being a law at all times (an eternal but time-dependent law) differs from (2)'s and (3)'s each being laws at different times (temporary laws). If there were no difference, then the laws' immutability would be trivial in that any apparent change in the laws (say, from (2) to (3)) would actually involve no change in the laws (but would instead involve a law like (1) holding at all times). By appealing to the laws' sub-nomic stability, I will argue that it is metaphysically impossible for the laws to change.[4] Of course, a philosopher should not take some respectable scientific theory and declare a priori that its truth is metaphysically impossible! However, I shall argue that talk of laws as "newly created"[5] while the universe cooled is a bad metaphysical gloss of a respectable scientific theory.

Let's start by looking at some unsuccessful ways to argue for the laws' immutability. Nineteenth-century enthusiasm for evolution led some natural philosophers to take seriously the possibility that the laws can change over time. Responding to these proposals, Henri Poincaré insisted that the laws cannot change. Rather, the laws entail that different regularities hold under different naturally possible conditions. Changes in those conditions should not be mistaken for changes in the laws. What has changed instead "are nothing but resultants" of the laws and accidental conditions; the genuine laws "remain intact." Poincaré's argument for this view seems to be that any change in the alleged "laws" must happen for some reason consisting partly of principles that remain unchanged in the transition: the genuine laws. They remain intact "since it will be through these principles that the changes will be made."[6]

The universe's early phase transitions may be best understood along these lines. The current "laws of low-energy physics" are then frozen accidents resulting from the fundamental laws together with a condition accidentally prevailing in our cosmic epoch: the state of the Higgs field(s). The current "laws of low-energy physics" were violated in the early universe because different accidental conditions prevailed then. Likewise, if the Higgs field(s) underwent some transition at 10^{-10} seconds after the Big Bang, then the laws governing that transition (by specifying the chance in those conditions that the Higgs field(s) would change to the state that has since prevailed) are genuine laws, along with laws specifying how particle interactions depend on the state of the Higgs field(s). None of them has changed.

However, Poincaré's argument (as I understand it) fails to show that the laws cannot change. Firstly, the argument presupposes that any alleged change in the laws must happen for some reason. But the fundamental laws are often taken to be brute facts (that is, facts that could have been otherwise, but there is no reason why they are not otherwise). That scientific explanations come to an end with the fundamental laws is what makes them fundamental. Just as there is no explanation of the fundamental laws, so presumably there would be no explanation of a change in the fundamental laws. It would be a brute fact that (2) is a law during one span of time and (3) is a law during another.

Secondly, even if Poincaré is correct in assuming that any alleged change in the laws must happen for some reason, why must the change be governed by a principle that remains unchanged in the transition? Here's an analogy. The Constitution codifies the fundamental laws of the United States. Article Five specifies the procedures for Constitutional amendment. An amendment could even amend Article Five. The ratification of the amendment would then be governed by Article Five, yet Article Five would not remain unchanged in the transition. Likewise, a change in the natural laws could happen for a reason, yet the laws governing the change could be among those that change.

Thirdly, even if Poincaré is correct in assuming that any alleged change in a law must be governed by a principle that remains unchanged in the transition, this constraint imposes no obstacle to that principle changing later, in accordance with some other principle that remains unchanged in that transition. No principle need be immutable even if some never actually change.

Now let's look at an argument aiming to derive the laws' immutability simply from their truth. Suppose for the sake of *reductio* that (2) is a law for some span of time and (3) is a law thereafter. Suppose that sometime during the latter period, two electrons have been at rest, separated by r centimeters, for at least r/c seconds. Then to accord with (3), these electrons must experience a mutual electrostatic repulsion of f dynes. But this occurrence violates (2). Since (2) is false, (2) is not a law. *Reductio* completed: the laws cannot change.

There are two points at which this argument should be resisted. First, (2) can cease to be a law, and (3) can thenceforward be a law, without violating the requirement that laws be truths—as long as (2) and (3) remain uninstantiated. Presumably, plenty of laws are uninstantiated (as we saw in chapter 2).[7] If accidentally there never are two electrons at rest exactly r centimeters apart for r/c seconds, then both (2) and (3) are true. So while (2) is a law, (3) must be an accident, and vice versa. Hence the *reductio* fails to preclude all changes in the laws: it permits vacuous truths to swap lawhood for accidenthood and vice versa.

Naturally, fans of mutable laws have something more dramatic in mind than that! But that the *reductio* leaves vacuous laws untouched highlights the fact that it appeals to nothing about laws beyond their truth. We might have expected the laws' immutability to derive instead from whatever distinguishes laws from accidents.

The second objection accuses the *reductio* of begging the question against the laws' mutability by presuming that if *m* is ever a law, then *m* is true. If the laws could change, then the laws of one period could presumably be violated during another period (when they aren't laws)—and in that event, they would not be true. To remain open-minded about whether there can be different laws during different periods, we should demand only that if *m* is a law throughout some period, then the events occurring *in that period* accord with *m*—in other words, that *m* be true "of that period," though perhaps not true *simpliciter* (that is, of the universe's entire history).

Let's say that *m* is true "of a given period" exactly when the universe's history during that period is logically consistent with *m*'s truth. Suppose that *m* is a law in a given period only if *m* is true of that period, but *m* need not be true *simpliciter*. Then the *reductio* fails: (2) can cease to be a law, and (3) can thenceforward be a law, even if during each period, there are electrons at rest separated by exactly *r* centimeters for at least r/c seconds.

Now let's see how the laws' stability bears upon the possibility of (2)'s being a law for the universe's first 10^{-10} seconds and (3)'s being a law thereafter.

3.3. Why the Laws Are Immutable

Laws form a sub-nomically stable set, and by definition, a sub-nomically stable set consists exclusively of truths. But as we just saw, this stipulation begs the question against the laws' mutability. Let's try to be more hospitable to the possibility of temporary laws by identifying the laws during a given period with the members of a set that is stable "for that period," where such a set's members may be false as long as they are true of that period.

Accordingly, for any logically closed, nonempty set of sub-nomic claims *m,* where each *m* is true of the given period, let the set qualify as "sub-nomically stable for that period" exactly when its members exhibit the kind of invariance under counterfactual suppositions that we found distinguishes laws from accidents—that is, exactly when all of the subjunctive conditionals demanded by chapter 1's definition of "sub-nomic stability" are true in every context. If some period's laws form a nonmaximal set that is sub-nomically stable for that period, then can the laws differ in different periods?

(Having dropped the requirement that a set's members be true *simpliciter* for the set to count as stable for a given period, should we lower the bar further by requiring their invariance only under those counterfactual suppositions that pertain exclusively to the given period? Later in this section, I shall consider this proposal.)

Suppose that (2) is a law while the universe is no more than 10^{-10} seconds old. Even if no two electrons actually are at rest for at least r/c seconds, exactly r centimeters apart, during the period when (2) is a law, (2) is not idle. Since (2) belongs to a nonmaximal set that is sub-nomically stable for that period, (2) specifies what *would have* happened then, *had* there been two such electrons: they would have experienced a mutual electrostatic repulsion of F dynes. The truth of

(4) Had two electrons been at rest and exactly r centimeters apart for at least r/c seconds at some moment when the universe is no more than 10^{-10} seconds old, then any such electrons would have experienced at that moment a mutual electrostatic repulsion of F dynes

is part of what makes a certain set containing (2) qualify as sub-nomically stable while the universe is no more than 10^{-10} seconds old. (I assume throughout that (4)'s antecedent is logically consistent with the relevant stable set, and likewise for my other examples.)

Suppose that 10^{-10} seconds after the Big Bang, (3) replaces (2) as a law. Even if no two electrons actually are at rest for at least r/c seconds, exactly r centimeters apart, during the period when (3) is a law, this conditional holds:

(5) Had two electrons been at rest and exactly r centimeters apart for at least r/c seconds at some moment when the universe is more than 10^{-10} seconds old, then any such electrons would have experienced at that moment a mutual electrostatic repulsion of f dynes.

(5)'s truth is part of what makes a certain set containing (3) qualify as sub-nomically stable for the period when the universe is more than 10^{-10} seconds old. There is no contradiction in both (4) and (5) being true.

However, here is another conditional that must be true for (2)'s set to count as sub-nomically stable for the pre-10^{-10}-second period:

> (6) Had two electrons been at rest and exactly r centimeters apart for at least r/c seconds at some moment when the universe is *more* than 10^{-10} seconds old, then any such electrons would have experienced at that moment a mutual electrostatic repulsion of F dynes.

But (5) and (6) cannot both be true!

The conditionals required for (2)'s lawhood during the earlier period conflict with the conditionals required for (3)'s lawhood during the later period. Unlike the *reductio* considered in the previous section, this argument for the laws' immutability exploits the laws' lawhood, not merely their truth. Furthermore, unlike that *reductio,* our new argument permits (2) and (3) to be uninstantiated, since it concerns subjunctive facts. Moreover, the argument allows m to be a law during a given period even if m is false, as long as m is true "of that period." Nevertheless, even after making these accommodations for temporary laws, we find that the laws in a given period must be laws forever.[8]

We can now see how (1)'s being a law at all times (an eternal but time-dependent law) would differ from (2)'s and (3)'s each being laws at different times (temporary laws). If the laws must form a sub-nomically stable set, then for (1) to be a law (at all times), (4) and (5) must be true, but (6) is not required, so there is no contradiction. In contrast, if (2) is a law during the earlier period and those laws must form a set that is stable for that period, then (4) and (6) must be true, which conflicts with the conditionals required for (3) to be a law during the later period.

Stability supplies another argument for the laws' immutability. In chapter 1, the nested counterfactuals in the definition of "stability" allowed us to conclude that a given stable set would still have been stable had q been the case, for any q that is logically consistent with the set. (To simplify matters slightly, I'll give the argument using would-conditionals rather than might-conditionals.) Suppose m is any member of stable set Γ, and $q, r, s \ldots$ are each logically consistent with Γ. Then q $\square\!\!\rightarrow m$, $q \square\!\!\rightarrow (r \square\!\!\rightarrow m)$, $q \square\!\!\rightarrow (s \square\!\!\rightarrow m)$, $q \square\!\!\rightarrow (r \square\!\!\rightarrow (s \square\!\!\rightarrow m))$, \ldots are all true. So in the closest q-world, m is true and $r \square\!\!\rightarrow m$, $s \square\!\!\rightarrow m$, $r \square\!\!\rightarrow (s \square\!\!\rightarrow m)$, \ldots are all true—which makes Γ stable in the closest q-world. Hence, if q is false, then had q been the case, the actual laws would still have been laws—presuming the members of a set that is stable *simpliciter* in a given world to be laws there. We thereby save a feature of scientific practice, such as our assent to "Had Earth's axis been nearly aligned with its orbital plane, then terrestrial seasons would have been quite different, though the actual laws of nature would still have been laws—which is *why* the seasons would have been so different." Now let's make a similar argument, but from Γ's stability *for a given period*. Again, $q \square\!\!\rightarrow m$, $q \square\!\!\rightarrow (r \square\!\!\rightarrow m)$, $q \square\!\!\rightarrow (s \square\!\!\rightarrow m)$, $q \square\!\!\rightarrow (r \square\!\!\rightarrow (s \square\!\!\rightarrow m))$, \ldots are all true. Suppose q is true. Then $q \square\!\!\rightarrow m$ is true only if m is true, and $q \square\!\!\rightarrow (r \square\!\!\rightarrow m)$ is true only if $(r \square\!\!\rightarrow m)$ is true, and so forth.[9] Hence m, $(r \square\!\!\rightarrow m)$, \ldots are true—so Γ is stable *simpliciter*! Therefore, Γ's members are laws forever, not merely during the given period. The laws are immutable.

By this argument, if Γ contains all and only the laws during a given period, then Γ's members are laws forever. Indeed, not only must all of Γ's members still be laws during a later period, but also no claim m that is not a law during the given period can be a law (along with Γ's members) during a later period—since it would then have to have been a law during the earlier period.

However, despite lowering the bar from stability *simpliciter* to stability "for a given period," we have still perhaps been insufficiently hospitable to the possibility of changing laws. The conditionals required for Γ's stability for a given period have turned out to ensure that Γ is stable *simpliciter*. Perhaps the only subjunctive conditionals $q \square\!\!\rightarrow m$, $q \square\!\!\rightarrow (r \square\!\!\rightarrow m)$, \ldots that we should have required, for Γ to qualify as stable for a given period, are those where $q, r, \ldots$ each concerns exclusively the given period and where m, a logical consequence of Γ, concerns exclusively the given period. (Let's say that p "concerns exclusively a given period"—say, when the universe is no more than 10^{-10} seconds old—exactly when necessarily, p is true if p is "true of the given period.") If lawhood during a given period is connected to this relaxed sense of stability for that period, then (2)'s lawhood for the period when the universe is no more than 10^{-10} seconds old appears *not* to demand that (6) be true, merely that (4) be true. The argument for the laws' immutability is blocked.

But during what period is (2) a law? Suppose that as a matter of accidental fact, the period when the universe is no more than 10^{-10} seconds old is exactly the period when the universe's temperature is not below 3×10^{15} K. So if (2)'s lawhood during this period is connected to (2)'s belonging to a nonmaximal set that is sub-nomically stable for this period (in the above, relaxed sense), then which conditional's truth does (2)'s lawhood demand, (7)'s or (8)'s?

> (7) Had two electrons been at rest and exactly r centimeters apart for at least r/c seconds at some moment when the universe is no more than 10^{-10} seconds old and is *below* 3×10^{15} K, then any such electrons would have experienced at that moment a mutual electrostatic repulsion of F dynes.

> (8) Had two electrons been at rest and exactly r centimeters apart for at least r/c seconds at some moment when the universe is not below 3×10^{15} K and is *more* than 10^{-10} seconds old, then any such electrons would have experienced at that moment a mutual electrostatic repulsion of F dynes.

If (7) is true and (8) is false, then (2)'s lawhood is set to expire when the universe's age exceeds 10^{-10} seconds, and this moment just happens to be when the universe's temperature falls below 3×10^{15} K. But then (1) is the genuine law, the laws never really change, and (2) was never a law. (Of course, since (1) is a law and logical consequences of laws are laws, it is a law that (2) is true of the period before the universe turns 10^{-10} seconds old.) Alternatively, if (8) is true and (7) is false, then (2)'s lawhood is set to expire when the universe's temperature falls below 3×10^{15} K. But then the genuine law is that between any two electrons that have been at rest, separated by r centimeters, for at least r/c seconds, there is an electrostatic repulsion of F dynes, if the universe is at least 3×10^{15} K, and f dynes ($f \neq F$) otherwise. Again, (2) was never a law.

Here is another way to put the same point. Suppose that conditionals like (7) are true whereas conditionals like (8) are false, so that (2) belongs to a nonmaximal set that is (in the relaxed sense) "sub-nomically stable for the period before the universe's age exceeds 10^{-10}

seconds"—and (3) likewise for the period thereafter. Then the conditionals whose truth makes these sets stable for those periods follow from the conditionals whose truth makes a set containing (1) stable *simpliciter*. So on this interpretation of the laws "changing," (2)'s being a law during the pre-10^{-10}-second period and (3)'s being a law thereafter adds nothing, as far as which conditionals hold, to (1)'s being a law forever. I suggest that the temporary laws add nothing *at all* here. Once the relevant period is designated as the period before the universe's age exceeds 10^{-10} seconds, the laws "changing" from (2) to (3) at that period's close is nothing but (1)'s being a law throughout the universe's history. We haven't got temporary laws; we've got an eternal (time-dependent) law.[10]

I conclude that any apparent change in the laws (say, from (2) to (3)) would actually involve no such thing, but instead some law (like (1)) holding eternally. Let's wrap up this section by contrasting my account to what two familiar philosophical accounts of lawhood say about the laws' (im)mutability.

According to Lewis's "Best System" account, the laws are the contingent generalizations in the deductive system (of truths about the Humean mosaic and claims about objective chances) having the optimal combination of simplicity, informativeness, and fit regarding the entire history of instantiations of all properties of an elite sort. It follows that the laws are immutable, since the laws at each moment are fixed in the same way by the same thing: the universe's complete history of elite-property instantiations.

However, Lewis's account entails the laws' immutability only because a certain adjustable parameter in the account has been set to "the universe's entire history." That parameter could be set differently. For example, there is a deductive system having the optimal combination of simplicity, informativeness, and fit regarding the elite-property instantiations *during a given period*. I see no grounds on which Lewis's account could object to deeming the members of that system to be the laws during that period. For example, Lewis's account is sometimes motivated as follows[11]:

YOU   Describe the universe please, Lord.

GOD   Right now, there's a particle in state Ψ_1 and another particle in state Ψ_2 and I'll get to the

	other particles in a moment, but in exactly 150 million years and 3 seconds, there will be a particle in state Ψ_3 and...
YOU (CHECKING WATCH)	Lord, I have an appointment in a few minutes.
GOD	All right, I'll describe the universe in the manner that is as brief and informative as it is possible simultaneously to be—by giving you the members of the "Best System."
YOU	Do tell...

But you might instead have begun the conversation by asking God to tell you about the goings-on during a given period of the universe's history. If what God ultimately gave you in the original imaginary conversation deserve to be considered "the natural laws," then by the same token, what God ultimately gives you in the second imaginary conversation merit being deemed "the laws during the given period."

I have argued that any "laws regarding the period before the universe was more than 10^{-10} seconds old" and "laws after the universe was 10^{-10} seconds old" presuppose eternal (though perhaps time-dependent) laws knitting them together, to the lawhood of which their "lawhood for those periods" adds nothing. The same cannot be said on Lewis's account. If p belongs to the Best System for the period before the universe was more than 10^{-10} seconds old and q belongs to the Best System for the period after the universe is 10^{-10} seconds old, then it does not follow that "p before the universe was more than 10^{-10} seconds old and q thereafter" (or another such fact, perhaps involving a critical temperature rather than time) is part of the Best System for the universe's entire history. For example, suppose that B-ons are a common species of elementary particle, but the first B-on comes into existence after the universe was 10^{-10} seconds old. Then "There are no B-ons" might well belong to the Best System for the period before the universe was more than 10^{-10} seconds old without "There are no B-ons before the universe was 10^{-10} seconds old" belonging to the Best System for the universe's entire history. Having a separate axiom concerning the universe's first 10^{-10} seconds might well decrease a system's simplicity by more than can be outweighed by the additional information it supplies, making any system with this axiom fall short of the Best for the universe's entire history. But "There are no B-ons" would not bring a similar disadvantage to a candidate for "Best System for the universe's first 10^{-10} seconds." Likewise, if "p before the universe was more than 10^{-10} seconds old and q thereafter" belongs to the Best System for the universe's entire history, then p need not belong to the Best System for the pre-10^{-10}-second period.

Indeed, if the laws of a given period are just the members of the Best System for that period, then the laws of March 2005 could in principle differ radically even from the laws of March 10, 2005. This welter of laws for different periods is at odds with scientific practice. Yet if their membership in the Best System for the universe's entire history is supposed to give certain facts various rights (for example, to heavily influence the truth-values of counterfactuals) and responsibilities (to help explain various events), then why wouldn't membership in the Best System for the period $[t_1, t_2]$ bestow some analogous significance on certain other facts?

Lewis's account avoids this result, rendering the laws immutable, only by restricting its attention to the Best System for the universe's entire history. But the rest of Lewis's account does not demand this restriction. The notion of "the Best System for the period $[t_1, t_2]$" is no less coherent than the notion of "the Best System for the universe's entire history." To fix the relevant period as the universe's entire history is artificial; it must be inserted by hand. If the laws are immutable, then Lewis's account contains an extra degree of freedom—a surplus adjustable parameter.

Of course, my argument for the laws' immutability depended crucially on certain views that Lewis rejects regarding the laws' relation to counterfactuals (namely, the laws' stability). So my argument for the laws' immutability did not proceed from neutral ground. Nevertheless, Lewis's account also entails that the laws are immutable—but only by the grace of an adjustable parameter's particular setting. The laws' immutability is dispensable rather than integral to the account.

Armstrong's account of laws as contingent relations among universals would seem better able to explain why the laws are immutable. Universals stand outside of the ebb and flow of particular events.[12] Presumably, then, the same can be said of their standing in certain relations; those facts cannot change. Accordingly, Armstrong initially argued that any relation among universals must hold omnitemporally.[13]

However, Armstrong now tends to think otherwise:

Why may it not be that F has the nomic relation [to] G at one time, but later, since the connection is contingent, this relation lapses, perhaps being succeeded by F's being related to H? . . . It seems that I have to allow that contingent relations between universals can change.[14]

If that is correct, then (since, I have argued, laws cannot change) laws cannot be "nomic–necessitation" relations among universals.

Can Armstrong's account really be disposed of so easily? I am inclined to think that the analysis of laws in terms of contingent "nomic-necessitation" relations among universals leaves unspecified whether the laws can change. The notion of a "nomic-necessitation relation" is underdescribed. Of course, the account could be made to stipulate that the nomic–necessitation relations holding among universals (such as Fness nomically necessitating Gness) support exactly the counterfactuals that are required for the corresponding set of truths (containing "All F's are G") to qualify as stable. Thus fortified, the account would (I have argued) entail the laws' immutability. But this tactic strikes me as building into the account expressly what the account should be explaining. Rather than getting the right answer by some ad hoc fine tuning added loosely to the core proposal, the account should offer an independent picture of what it is for universals to stand in nomic-necessitation relations, and from this picture, the laws' immutability should fall out naturally.[15]

3.4. Symmetry Principles as Meta-laws

In this section and the next, I turn to a new topic: "symmetry principles" and what it would take for them to be "meta-laws" (that is, laws that govern the laws governing sub-nomic facts). Let's work our way toward our target by starting with some of the symmetries exhibited by Coulomb's law (C):

For any time t, any locations r_1 and r_2, and any quantities q_1 and q_2 of electric charge, two point bodies that have been at those locations with those quantities of charge from time $(t - |r_1 - r_2|/c)$

until time t exert on each other at t an electrically repulsive force $F = q_1 q_2 / |r_1 - r_2|^2$.

According to C, two bodies that have long been at rest, both charged to $+5$ statcoulombs, exert the same mutual electrostatic force as two bodies long at rest, separated by the same distance as the first pair, but both charged to -5 statcoulombs. Indeed, if we take a world where C holds and reverse the signs of all charges there, but leave unchanged everything else (including the absolute values of those charges, their distances apart, and their mutual electrostatic forces), then the resulting world still accords with C. (This is not a contingent truth expressed by a counterfactual conditional. Rather, it is a logical truth expressed by an indicative conditional—a purely formal feature of C.) Instead of expressing this point in terms of hypothetical changes to the world, we could put it in terms of hypothetical changes to the law: C is unchanged if q_1 is replaced by $-q_1$ and q_2 is replaced by $-q_2$:

$$F = q_1 q_2 / |r_1 - r_2|^2 \rightarrow F = (-q_1)(-q_2)/|r_1 - r_2|^2$$
$$= q_1 q_2 / |r_1 - r_2|^2.$$

A law is "symmetric" in a certain respect exactly when it remains unchanged under a certain transformation.

Coulomb's law is also symmetric under arbitrary spatial displacement—that is, under a shift of all bodies' positions at all times by the same vector a. Transformed by $r_i \rightarrow r_i + a$, C becomes $F = q_1 q_2 / |[r_1 + a] - [r_2 + a]|^2 = q_1 q_2 / |r_1 - r_2|^2$. The law doesn't care about the bodies' absolute positions; only their separation matters. Coulomb's law is also time-displacement symmetric: unchanged under the transformation $t \rightarrow t + a$ for arbitrary temporal interval a.

These symmetries are not confined to Coulomb's law. The laws of fundamental physics as a whole are thought to display space-displacement and time-displacement symmetries along with symmetry under rotations and under velocity boosts—expressing the irrelevance of absolute times, positions, directions, and steady straight-line motions.[16] Presumably, there might have been a law privileging, say, a given point in space. For example, it might have been a law that each body at any nonzero distance r from the universe's center feels a force q^2/r^2 toward the center. This law is not invariant under arbitrary spatial displacement. The result of taking a possible world that (nonvacuously) accords with this law and shifting all of its bodies by *a,* but leaving their charges and the forces they feel (along with the location of the universe's center) unchanged, is a world that fails to accord with this law.

Generalizing from one symmetry exhibited by one law, a "symmetry principle" ascribes some symmetry to the laws as a whole. In stating a symmetry principle, we must be careful since every logical consequence of laws is a law. Hence, if C is a law, then for a given moment T in the universe's history, it is a law that for any two point bodies at any time $t > T$, $F = q_1 q_2 / |r_1 - r_2|^2$. This consequence of C pertaining only to times $t > T$ is not time-displacement symmetric; a world can satisfy it while departing from Coulomb's law before T, but if all events are then shifted by some interval *a* into the future, an exception to Coulomb's law may get shifted to a time after T, thereby violating this consequence of C. Nevertheless, the principle of time-displacement symmetry should not preclude this consequence of C from being a law—on pain of precluding any world governed by Coulomb's law! Accordingly, the symmetry principle should require merely that every law *follow* from time-displacement symmetric laws—in other words, that the laws *as a whole* be unchanged under arbitrary time-displacement.[17]

My earlier example of a law privileging the universe's center suggests that symmetry principles are not logically, conceptually, or metaphysically necessary.[18] If a given symmetry principle holds, then it follows logically from the facts about which sub-nomic facts are laws and which are not. A symmetry principle captures a *regularity* among the laws: that every law is unchanged under a certain transformation (or follows from laws that are). What is this regularity's modal status?

Let's turn momentarily from this regularity among the laws to a regularity among the sub-nomic facts. Such a regularity is either accidental or a law. Since C is a law, it is no coincidence that the members of every pair of point charges exert mutual electrostatic forces in accordance with the Coulombic equation. This regularity is not a consequence of the circumstances in which the actual pairs of point charges happen to find themselves. Rather, the regularity would still have held even if those circumstances had been different. The regularity *has* to hold. That C is a law explains the electrostatic forces that various charges actually exert on each other. C's lawhood is explanatorily prior to the facts that C governs. In contrast, that all gold cubes are smaller than one cubic mile holds as a byproduct of the sizes that the actual gold cubes happen to have. The reason why all of the pairs of point charges in one space-time region exert mutual electrostatic forces in accordance with the Coulombic equation is the same as the reason why all of the pairs of point charges in another space-time region do so. But the reasons why all of the gold cubes in one space-time region are smaller than one cubic mile may have nothing in common with the reasons why all of the gold cubes in another space-time region are smaller than one cubic mile.

An analogous distinction should apply to a regularity *among the laws*. A given symmetry principle may be a meta-law (that is, a "second-order law") governing the first-order laws—a requirement to which the laws that govern sub-nomic facts *must* adhere. Otherwise, it is a byproduct of the first-order laws—their offspring, as it were, holding merely in virtue of what those laws happen to be.[19] If time-displacement symmetry is a meta-law, then it explains why every first-order law is time-displacement symmetric (or follows from ones that are)—just as the fact that C is a law explains why the members of every pair of point charges exert mutual forces in accordance with a certain equation. In contrast, if time-displacement symmetry is a byproduct of the first-order laws, then it is not explanatorily prior to those laws (just as the gold-cubes regularity holds only because this gold cube is smaller than one cubic mile, and that one is too, and so forth). In that case, the invariance under time-displacement of the law governing one fundamental force has no explanation in common with the invariance under that transformation of the law governing another fundamental force. That every first-order law is invariant under that transformation (or follows from ones that are) is just a giant coincidence. As Nobel physics laureate C. N. Yang remarked, one way to interpret symmetry principles is as "only consequences of the dynamical laws that by chance possess the symmetries."[20]

Obviously, "by chance" here does not mean "resulting from some chance set-up" or even "by happenstance." Though a symmetry principle can be a byproduct of the first-order laws, it cannot be accidental in precisely the manner of all gold cubes being smaller than a cubic mile. A symmetry principle's truth is a consequence solely of the facts about which sub-nomic claims *m* are laws and which are not, whereas an accident does not follow solely from these facts. Every symmetry principle (whether meta-law or byproduct of the first-order laws) holds in every possible world with exactly the same facts of the form

"It is a law that *m*" (where *m* is sub-nomic) as in the actual world. Plainly, the same cannot be said of accidental truths. Thus, whereas a first-order regularity's lawhood is equivalent to its being non-accidental, a symmetry principle's meta-lawhood is not equivalent to its being non-accidental. (Consequently, I have called symmetry principles that are not meta-laws "byproducts" rather than "accidents.")

Accordingly, we have some further work to do in order to understand what it would be for a symmetry principle to be a meta-law. You might have thought that a meta-law was simply a law about other laws— differing from first-order laws only in its content, not in its modal status. However, this is not so: the first-order laws are simply the nonaccidents among the sub-nomic truths, but every symmetry principle is nonaccidental, whether it is a meta-law or a byproduct. I will suggest that a meta-law differs from a byproduct of the first-order laws by possessing a stronger variety of natural necessity than first-order laws possess. That is ultimately why meta-laws are explanatorily prior to first-order laws.

(Did you pause earlier when you read that a symmetry principle that is a byproduct of the first-order laws holds merely in virtue of what those laws *happen to be*? Despite the natural necessity of the first-order laws, they typically lack the exalted modal status of a meta-law. Compared to a meta-law, the first-order laws—at least, those that fail to follow exclusively from meta-laws—merely happen to hold.)

If a symmetry principle is a byproduct of the first-order laws, then the fact that every first-order law is invariant under a certain transformation (or follows from first-order laws that are) holds "by chance": as nothing more than a formal peculiarity of the first-order laws. According to Nobel physics laureate Steven Weinberg, various elementary-particle symmetries, such as isospin symmetry (roughly speaking, invariance under the transformation of protons into neutrons and vice versa), were once regarded as "mathematical tricks; the real business of physicists was to work out the dynamical details of the forces we observe."[21] In other words, all of the explanatory work was once thought to be done by the first-order laws, which (roughly speaking) specify the forces on a system's components given the system's current configuration and how those forces, in turn, relate to the system's subsequent behavior. Symmetry principles like isospin symmetry were believed to have no explanatory significance. However, Weinberg says, nowadays such a symmetry is considered "as a fundamental fact about nuclear physics that stands on its own, independent of any detailed theory of nuclear forces." This is an ontological point: rather than being indebted to the first-order laws to make it true, the symmetry principle is prior to those laws. It is a requirement on any "detailed theory of nuclear forces": not merely an epistemic constraint on any theory worth taking seriously, but a "fundamental fact"—an ontological constraint on the nuclear forces themselves. As Nobel physics laureate Eugene Wigner says, such symmetry principles are laws "which the [first-order] laws of nature have to obey."[22] In virtue of their modal status, symmetry principles explain why the first-order laws possess certain features.[23]

Space-time symmetries, such as invariance under arbitrary spatial and temporal displacements, were considered meta-laws long before it was common to so regard non-space-time symmetries, such as isospin symmetry. Indeed, Lagrange and Hamilton appealed to space-time symmetry principles to explain why various conservation laws hold.[24] This kind of explanation is nearly universally accepted today: it is widely believed that time-displacement invariance explains why energy conservation holds, space-displacement invariance explains why linear momentum conservation holds, rotational invariance explains why angular momentum conservation holds, and symmetry under velocity boosts explains why the velocity of the center of mass is conserved.[25] Given the fundamental dynamical law (relating the system's behavior to its kinetic and potential energies in various configurations) and certain other conditions, each of these symmetry principles entails the corresponding conservation law.

But by the same token, given the fundamental dynamical law and those conditions, each of these conservation laws entails the corresponding symmetry principle.[26] Why, then, does the symmetry principle help to explain the conservation law, rather than the reverse? Presumably, the answer is that the symmetry principle is a meta-law and so governs the various first-order laws, including the conservation law. The symmetry principle has greater modal force than the conservation law and so can explain it, but the conservation law lacks the symmetry principle's modal force and so cannot explain it. Hence, to understand why certain symmetry principles are explanatorily prior to various conservation laws, we must understand what it is for a symmetry principle to be a law governing other laws.

I am not trying to show that there actually are meta-laws or that a particular symmetry principle is among them. Those are matters for science to discover. Indeed, some of the most important discoveries of physics are generally treated as meta-laws. Foremost among these is this meta-law at the heart of Einstein's special theory of relativity:

> General laws of nature are covariant with respect to Lorentz transformations. This is a definite mathematical condition that the theory of relativity *demands* of a natural law....[27]

> The content of the [special] relativity theory can . . . be summarized in one sentence: all natural laws *must* be so conditioned that they are covariant with respect to Lorentz transformations.[28]

My aim is to understand this "demand"—this "must"—by identifying what it would be for a symmetry principle to be a meta-law rather than a byproduct of the first-order laws.

Closely related to the Lorentz-covariance meta-law is the "principle of relativity": that in all inertial reference frames, the fundamental first-order laws are the same. Roger Penrose characterizes Einstein's insight as "that one should take relativity as a *principle*"—a meta-law—"rather than as a seemingly accidental consequence of other laws."[29] Our goal in the next section will be to understand how some truths about the first-order laws hold as meta-laws whereas others, as byproducts of the first-order laws, could aptly be termed "accidental" (despite following exclusively from facts about the laws).

A philosophical account of natural law should portray the meta-laws' relation to the laws they govern as strictly analogous to the first-order laws' relation to the (sub-nomic) facts they govern.[30] I shall now explain how my account does so.

3.5. The Symmetry Meta-laws Form a Nomically Stable Set

Just as the key difference between first-order laws and accidents involves a law's invariance, had the sub-nomic facts been different, so the key difference between meta-laws and byproducts of the first-order laws involves a meta-law's invariance, had the first-order laws been different.

Meta-laws persist under any counterfactual supposition concerning the first-order laws that is logically consistent with the meta-laws. As we will now see, the meta-laws thereby form a set Λ^{meta} possessing *nomic* stability just as the first-order laws form a set Λ possessing *subnomic* stability.

For example, if time-displacement symmetry is a byproduct of the particular kinds of forces there actually are, then had the details of the force laws been different, the symmetry might not still have held—just as had the gold cubes' sizes been different or had there been additional gold cubes, then some might have been larger than a cubic mile. But as a meta-law, the symmetry would still have held, had the force laws been different. No moment of time would have been privileged even if gravity had been different or there had been additional, alien forces alongside gravity, electromagnetism, and so forth.

For a first-order law such as "All emeralds are green," our past observations of emeralds confirm not only that all of the actual emeralds lying undiscovered in some far-off land are green, but also that my pocket would have contained something green, had an emerald been in my pocket now. Likewise, scientists sometimes regard the symmetry of known force laws as confirming that the same symmetry principles hold of whatever unknown laws govern as yet undiscovered kinds of forces.[31] As in the emerald example, this confirmation fails to discriminate between actual unexamined cases and counterfactual cases; the evidence confirms that had there been an additional or different kind of force, its laws would have exhibited the same symmetry. Evidence bears on the forces there would have been for the same reason as it bears on the unknown forces there are. Accordingly, as Wigner says, symmetries that are meta-laws are not "based on the existence of specific types of interaction."[32]

That symmetry meta-laws transcend those specifics is reminiscent of what we learned in chapter 1 regarding conservation laws. There we found that conservation laws (along with certain other laws, such as the law of the composition of forces and the fundamental dynamical law) may form a sub-nomically stable, proper subset of Λ that omits the force laws, since plausibly the conservation laws would still have held even had the force laws been different. Just as scientists regard our evidence for the symmetry meta-laws as confirming various counterfactuals, so likewise scientists have generally taken their evidence for

the conservation laws (and their colleagues in a stable proper subset of
Λ) not merely as confirming that any actual unknown phenomenon,
even one involving as yet unfamiliar kinds of forces, accords with
energy and momentum conservation, but also as confirming that
energy and momentum would still have been conserved, had there
been additional or different kinds of forces.[33]

Accordingly, symmetry principles and conservation laws are
often grouped together as standing loftily above the piddling idiosyncrasies of the various force laws. As Nobel physics laureate Richard Feynman says,

> When learning about the laws of physics you find that there are
> a large number of complicated and detailed laws, laws of gravita-
> tion, of electricity and magnetism, nuclear interactions, and so
> on, but across the variety of these detailed laws there sweep great
> general principles which all the laws seem to follow. Examples of
> these are the principles of conservation, certain qualities of sym-
> metry, the general form of quantum mechanical principles.... [34]

Conservation laws that are explained by symmetry principles cannot
be coincidences of the particular kinds of forces there happen to be,
but rather must belong to a sub-nomically stable proper subset of Λ. As
Wigner says,

> [F]or those [conservation laws] which derive from the geometri-
> cal principles of invariance [that is, the space-time symmetries—
> ML] it is clear that their validity transcends that of any special
> theory—gravitational, electromagnetic, etc.—which are only
> loosely connected.[35]

As we saw in chapter 2, the natural necessity possessed by these conservation laws is stronger than the kind possessed by the "special theories"
of the various forces; the force laws belong to Λ but not to the sub-
nomically stable proper subset of Λ to which these conservation laws
belong. Now we will see that the reason why these conservation laws
belong to this sub-nomically stable proper subset of Λ and so possess
this stronger variety of natural necessity is that the symmetry principles
responsible for them are meta-laws. Possessing an even stronger kind of
natural necessity, they are empowered to bestow a strong kind of natural necessity upon the conservation laws.

To explain why energy is always conserved, we could derive its conservation from the fundamental dynamical law, the various particular force laws, and the fact that these force laws exhaust the actual kinds of forces. This derivation even explains why energy conservation *is a law*, since all of the premises are matters of law. However, this explanation fails to explain why energy conservation *transcends* the details of the force laws—that is, why energy conservation is located above the force laws in the pyramid of first-order laws (fig. 1.3). In other words, this explanation fails to account for the conservation law's membership in a sub-nomically stable set that omits the force laws. With their weaker species of natural necessity, the force laws lack sufficient modal weight to explain why the conservation laws possess their stronger variety of natural necessity.

Likewise, the variety of necessity possessed by symmetry meta-laws is stronger than the variety possessed by conservation laws, empowering the former to explain the latter and precluding the reverse. The regularity associated with a symmetry principle is that each sub-nomic truth *m where it is a law* that *m* is invariant under a certain transformation (or follows from such truths that are)—a regularity in the regularities *associated with laws* governing sub-nomic facts.[36] A symmetry principle does not demand something of every fact, only of the laws; it states a fact about the laws governing sub-nomic facts. Hence, a symmetry principle is *not* a sub-nomic fact. It is ineligible for membership in Λ or in any other set possessing sub-nomic stability. It is "meta" to those laws. In contrast, the regularity associated with a conservation law (namely, that a given quantity is conserved) does not concern laws. It is a sub-nomic fact, belonging to Λ.

A truth that is "meta" to the laws in Λ (what I shall term a "nomic" fact) is a "meta-law" exactly when it *requires* something of the first-order laws; it is a constraint on rather than a byproduct of them. To elaborate this notion, we must take the relation between laws in Λ and the sub-nomic facts they govern and replicate it one level higher—at the meta level. Let's begin by turning from the sub-nomic claims (such as "All emeralds are green") alone to the sub-nomic claims together with the "nomic" claims (such as "It is *a law* that all emeralds are green" and "It is *not a law* that all gold cubes are smaller than a cubic mile"). Let a claim be "nomic" exactly when it purports to state a fact about which sub-nomic facts are matters of law—but not about (hold on now!) which of the facts about which sub-nomic facts are matters of law are themselves matters of law. (Let the truths among the nomic claims be the "nomic facts.") For example, that all first-order laws are time-displacement symmetric (or follow from ones that are) is a nomic fact. But the fact that *it is a law that* all first-order laws are time-displacement symmetric (or follow from ones that are)—that's one level higher even than the nomic facts.

That is, a claim is "nomic" exactly when, in every possible world, it is true or false wholly in virtue of (i) the sub-nomic facts there and (ii) which of them are laws and which are not, and (to exclude the sub-nomic claims) in some possible world, it is true or false at least partly in virtue of (ii). To define *nomic stability*—the nomic analogue of sub-nomic stability—we can simply take chapter 1's definition of "sub-nomic stability" and replace every appearance of "sub-nomic" there with "nomic or sub-nomic." (For the remainder of this section only, let's suspend our policy of reserving lower-case italicized English letters for sub-nomic claims.) The result:

> Consider a nonempty set Γ of truths that are nomic or sub-nomic containing every nomic or sub-nomic logical consequence of its members. Γ possesses *nomic stability* if and only if for each member m of Γ (and in every conversational context),
> $$\sim (p \lozenge\!\!\rightarrow \sim m),$$
> $$\sim (q \lozenge\!\!\rightarrow (p \lozenge\!\!\rightarrow \sim m)),$$
> $$\sim (r \lozenge\!\!\rightarrow (q \lozenge\!\!\rightarrow (p \lozenge\!\!\rightarrow \sim m))), \ldots$$
> for any nomic or sub-nomic claims $p, q, r, \ldots$ where $\Gamma \cup \{p\}$ is logically consistent, $\Gamma \cup \{q\}$ is logically consistent, $\Gamma \cup \{r\}$ is logically consistent, and so forth.

Roughly speaking, a closed set of truths that are nomic or sub-nomic qualifies as "nomically stable" exactly when (whatever the conversational context) the set's members would all still have held (indeed, none of their negations might have held) under every nomic or sub-nomic supposition logically consistent with the set—however many such suppositions are nested. A nomically stable set's members are as collectively invariant under nomic and sub-nomic suppositions as they could collectively be. They possess a kind of maximal invariance and thereby earn a genuine species of (natural) necessity. They are inevitable in that whatever might have happened (nomically or sub-nomically speaking), none of them is such that it might have been false (and *that,* in turn, is true whatever might have happened, and so forth).

What is the relation between a nomically stable set and the sub-nomically stable sets? For any nomically stable set, its sub-nomic members form a sub-nomically stable set. To show this, notice first that if p (a sub-nomic claim) is logically inconsistent with a nomically stable set Γ, then Γ must entail $\sim p$ (also sub-nomic), and so p is logically inconsistent with the set Σ containing exactly Γ's sub-nomic logical consequences. Conversely, if p is logically inconsistent with Σ, then obviously p is logically inconsistent with Γ. By Γ's nomic stability, Σ is preserved under any sub-nomic antecedent p that is logically consistent with Γ—which (we have just seen) are exactly those that are logically consistent with Σ. Hence, Σ is sub-nomically stable. Therefore, any sub-nomic truth that follows from a meta-law belongs to a sub-nomically stable set—one higher on the pyramid than Λ (since not all first-order laws follow from meta-laws). In other words, any sub-nomic truth that is a constraint imposed by the meta-laws on the first-order laws is itself a first-order law. Thus the meta-laws constrain the first-order laws.[37]

Λ lacks nomic stability. If it is a law that m (where m is sub-nomic), then "Had m not been a law, m would still have held" is typically false. (For example, it is not the case that energy would still have been conserved had there been no legal obligation for it to be.) This counterfactual's antecedent is a nomic claim logically consistent with Λ, so Λ's nomic stability requires this conditional's truth.

Instead take the set Λ^+ specifying which sub-nomic claims are laws and which are not. That it is a law that all emeralds are green belongs to Λ^+, and that it is not a law that all gold cubes are smaller than a cubic mile also belongs to Λ^+. The sorts of examples we examined in chapter 1 suggest that Λ^+ possesses nomic stability. For example, had Earth's axis been nearly aligned with its orbital plane, then terrestrial seasons would have been quite different, though the actual laws of nature would still have been laws—which is *why* terrestrial seasons would have been so different.[38]

Both meta-laws and byproducts of the first-order laws belong to Λ^+; the set includes all of the regularities in the first-order laws. Meta-laws are distinguished by their membership in more exclusive nomically stable sets. For familiar reasons, the nomically stable sets must form a pyramidal hierarchy (see fig. 3.1). Are there any other good

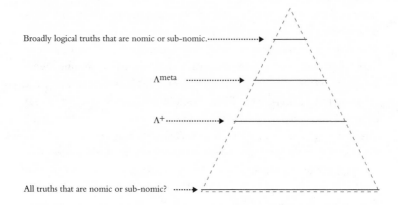

Broadly logical truths that are nomic or sub-nomic.

Λ^{meta}

Λ^{+}

All truths that are nomic or sub-nomic?

Figure 3.1   Some (though perhaps not all) good candidates for nomically stable sets.

candidates for nomic stability, located in the pyramid above Λ^{+} and below the set of broadly logical truths? Yes: a set Λ^{meta} containing various symmetry principles (and perhaps certain other nomic facts as well) but without the force laws, the fundamental dynamical law, the conservation laws, and other such first-order laws. (Λ^{meta} omits both the sub-nomic fact "Any two point bodies exert electrostatic forces..." and the corresponding nomic fact "It is a law that any two point bodies....") For a nomic fact to be a "meta-law" is for it to belong to such a set.

The nomic stability of Λ^{meta} requires that its members would still have held, had the force laws governing the actual kinds of forces been different, or had there been additional kinds of forces besides the actual kinds, or had the fundamental dynamical law been different, or had there been other such differences in the first-order laws. Of course, the symmetry meta-laws would not still have held, had there been an additional kind of force and a fundamental dynamical law that together entail that a given symmetry meta-law is false. But that supposition is logically inconsistent with Λ^{meta}, so Λ^{meta}'s nomic stability does not require Λ^{meta}'s preservation under that supposition.

What other meta-laws might there be, joining symmetry principles in Λ^{meta}? In a deterministic world, determinism might be just a byproduct of the first-order laws. Those laws might just happen to be rich enough to determine the future given the past. On the other hand, determinism there might be no coincidence. A meta-law might require that the first-order laws be sufficiently rich to make the universe deterministic. In that event, had the force laws been different, determinism would still have held. I will return to this point in chapter 4.

Suppose various symmetry principles to be meta-laws. The associated conservation laws neither join them to form a nomically stable set nor form a nomically stable set themselves. That is because had the fundamental dynamical law been different, the symmetries would still have held but the conservation laws need not have. For example, had the fundamental dynamical law been that F is proportional to mv rather than to ma, then the force laws and other first-order laws would still have been symmetric under arbitrary spatial and temporal displacements—but energy and angular momentum would not have been conserved.[39] As another example, suppose that the "force majeure law" had held: when two bodies of unequal mass collide, the less massive one disappears and the more massive one continues moving as it would have had no collision taken place. The familiar symmetries would still have held, but not the conservation laws. Notice that the force majeure law is logically consistent with the symmetry principles and associated conservation laws. If it is a law that there is only one body in the universe's entire history, that its mass remains unchanged, and that it moves uniformly forever, then the force majeure law holds (vacuously) and it is also a law that energy, momentum, and so forth are conserved. So the conservation laws *could* still have been laws under the supposition of the force majeure law. But they *wouldn't* still have been. Rather, there would still have been many bodies in the universe. So the conservation laws (even together with the symmetry principles) fail to form a nomically stable set.[40]

In contrast, if symmetries are meta-laws, then Λ^{meta} is nomically stable. A counterfactual supposition under which the symmetry principles fail to be preserved must be logically inconsistent with the meta-laws. But the conservation laws fail to be preserved under some nomic suppositions with which they are together logically consistent.

Even if the conservation laws (along with certain other laws, such as the fundamental dynamical law and the parallelogram of forces), without the force laws, form a proper subset of Λ that is subnomically stable, the range of suppositions under which these laws would still have held, by their sub-nomic stability, is narrower than the range of suppositions under which the symmetry meta-laws would still have held, by their nomic stability. (For example, unlike these conservation laws, symmetry meta-laws would still have held, had the fundamental dynamical law been different.) Thus, symmetry meta-laws have a stronger variety of natural necessity than conservation laws. So these conservation laws cannot help to explain the corresponding symmetry principles, since the range of suppositions under which these conservation laws are preserved is too narrow to give the symmetry principles their broader range of perseverance as meta-laws. A conservation law, even if it belongs to the sub-nomically stable proper subset of Λ, cannot explain why the corresponding symmetry would still have held, had the fundamental dynamical law been different.

Of course, just as there are some suppositions under which the symmetry principles are preserved but the conservation laws are not, so there are also some suppositions under which the conservation laws are preserved but the symmetry principles are not—for example, "Had it been a nonvacuous law that each body always moves at 5 meters per second in the $+x$ direction." It is *not* the case that symmetry meta-laws are invariant under a broader range of counterfactual suppositions than the associated conservation laws, just as it is not the case that a first-order law is invariant under a broader range of counterfactual suppositions than an accident (as I showed in chapter 1, section 1.4). Rather, the range of suppositions under which symmetry principles are invariant *in virtue of which they possess a species of necessity* (namely, in connection with Λ^{meta}'s nomic stability) is broader than the range of suppositions under which conservation laws are invariant *in virtue of which they possess a species of necessity* (namely, in connection with the sub-nomic stability of a proper subset of Λ). The supposition positing a nonvacuous law that each body moves always at 5 m/s in the $+x$ direction is logically inconsistent with Λ^{meta} (in particular, with rotational symmetry) and so falls outside the range of suppositions under which Λ^{meta} must be preserved to qualify as nomically stable.[41] In contrast, there are (as we have seen) nomic suppositions logically consistent with the set of conservation laws but under which the conservation laws are not all preserved as laws, thereby failing to qualify as nomically stable. Hence, there is no "symmetry" (!) between the symmetry principles and conservation laws, even though each is preserved under some suppositions under which the other is not.

I have emphasized not the range of counterfactual suppositions under which the set of symmetry principles would still have held, but instead the range under which they would still have held *by the set's nomic stability*. I have done so because (as I suggested in chapter 2) a set's stability involves its members together being as invariant as they could together be, and so (unlike just any old broad range of invariance) is associated with a variety of necessity. A law possesses its distinctive explanatory power by virtue of its necessity: p's lawhood explains why p obtains by rendering p inevitable, unavoidable— necessary. (I will say more about this in chapter 4.) Symmetry principles are explanatorily prior to conservation laws because they possess a stronger species of necessity (associated with their nomic stability) than the conservation laws do.

Let's sum up. When a conservation law is explained by a symmetry principle, the symmetry principle functions as the "covering law" and the "initial conditions" consist of the fundamental dynamical law (plus the other conditions required for the space-time symmetries to be associated with the familiar conservation laws; see note 26). The dynamical law is governed by the symmetry principle; the symmetry would still have held even if the dynamical law had not. The symmetry principle, in belonging to a nonmaximal nomically stable set, possesses a variety of necessity. It explains the conservation law by making it likewise necessary that if the dynamical law (and other conditions) hold, the conservation laws hold. Therefore, the conservation law that is explained by a symmetry principle transcends the particularities of the force laws: it joins the fundamental dynamical law in a subnomically stable set that does not include the force laws. That the conservation law would still have held, had the force laws been different, follows from the fact that not only the fundamental dynamical law, but also the symmetry principles would still have held, had the force laws been different—since the symmetry principles are meta-laws.[42]

Thus, symmetry principles are explanatorily prior to conservation laws when symmetry principles are meta-laws. I have elaborated meta-lawhood in terms of nomic stability. The meta-laws' relation to the laws they govern thereby mirrors the first-order laws' relation to the facts they govern.[43]

3.6. The Relation between Chancy Facts and Deterministic Laws

I'll now turn to this chapter's final topic. Suppose that unobtanium-346 is a rare radioactive isotope; only a few such atoms ever exist.[44] Suppose that each [346]Un atom, upon coming into existence, has a half-life of 7 microseconds: it possesses a 50% chance of decaying within the next 7 microseconds (and, perforce, a 50% chance of lasting longer than 7 microseconds). This is an *objective* chance, not a measure of my own personal degree of confidence. Suppose further that in fact, each of the few [346]Un atoms that ever exists decays within 7 microseconds of its creation. So we have two distinct facts:

> (1) Every [346]Un atom decays within 7 μs of its creation.
> (50%) Every [346]Un atom, at its creation, has a 50% chance of decaying within 7μs.

Although (1) and (50%) can be true together, they plainly cannot together be laws of nature unless it is a law that there are no [346]Un atoms. Indeed, the nonvacuous mere *truth* of (50%) logically precludes (1)'s lawhood. If (50%) is true nonvacuously, then it is possible for a [346]Un atom to exist and every [346]Un atom to last longer than 7μs. But if (1) is a law, then it is impossible for a [346]Un atom to exist and every [346]Un atom to last longer than 7μs. Whereas (50%)'s nonvacuous truth logically precludes (1)'s lawhood, the phenomenon is not symmetrical: (1)'s nonvacuous truth is logically compatible with (50%)'s lawhood. If (50%) is a statistical law governing [346]Un decay and the laws permit a [346]Un atom to exist, then (1) can be an accidental truth but cannot be a law.

The italicized entries in figure 3.2 form the asymmetry in which I am interested. Why do chancy facts, such as (50%)'s nonvacuous truth, preclude deterministic laws such as (1), while (1)'s nonvacuous truth fails to preclude (50%)'s lawhood?

This relation between chancy facts and deterministic laws seems important. I'll now look at how Lewis's account explains it so that in the next section, we may compare this explanation to mine.

Lewis says that only a system that respects this relation between chancy facts and deterministic laws is eligible for the competition for

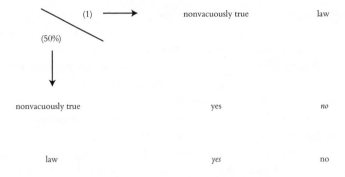

Figure 3.2   The relation between chancy facts and deterministic laws:
the two possible combinations are marked "yes" and the two impossible
combinations are marked "no."

"Best System." To enter the competition, a system must satisfy what
Lewis calls a "requirement of coherence," which says that "these systems
aren't in the business of guessing the outcomes of what, by their own
lights, are chance events: they never say A without also saying that A
never had any chance of not coming about."[45]

Here is the way I think this requirement is supposed to work. If (1)
is a law, then (1) belongs to the Best System, and by the requirement of
coherence, no system containing (1) can enter the competition for Best
System unless it says that (1) never had any chance of not coming
about. That is, regarding any circumstance in which a ^{346}Un atom could
exist without contradicting the system, the system must say that a ^{346}Un
atom in that circumstance never had any chance of lasting beyond 7μs.
So if (1) is a law, then it must be true that no ^{346}Un atom ever has any
chance of lasting beyond 7μs. But if (50%) is nonvacuously true, then
there exists a ^{346}Un atom that, at its creation, has a 50% chance of lasting
beyond 7μs. Thus (50%)'s nonvacuous truth logically precludes (1)'s
lawhood.[46]

The "requirement of coherence" gives the correct result. But is its
success achieved merely by its building the desired relation between
chancy facts and deterministic laws directly into Lewis's account? On
that account, laws and chances arise together from the global mosaic of
elite-property instantiations (see chapter 2, section 2.3). The chances
are not metaphysically independent of the laws. Rather, the fact that a

given chance holds at a given moment is nothing more than a fact about the laws and the history of elite-property instantiations through that moment—namely, that the history and laws entail that the chance holds. If facts about chances, unlike facts about elite-property instantiations, all obtain in virtue of the laws, then a requirement that laws and chances "cohere" may not seem ad hoc. Chancy facts must already be coordinated with the laws simply because of what chancy facts are. That (1)'s lawhood (but not (1)'s nonvacuous truth) interferes with (50%)'s nonvacuous truth might then be expected.

However, to motivate Lewis's "requirement of coherence," it does not suffice merely to posit the law-dependence of chances: that (50%) is true only if there are laws specifying that any ^{346}Un atom under certain elite-property conditions (that all actual ^{346}Un atoms in fact occupy) has a 50% chance of decaying within 7μs. The "requirement of coherence" goes significantly further: it demands that any such law preclude (1)'s nonvacuous lawhood. This feature of lawhood might have been expected to fall nicely out of Lewis's account—to result automatically from the competition for Best System, just as the existence of certain uninstantiated laws does (as we saw in chapter 2). But Lewis inserts the relation by hand, adding to the competition's rules a constraint on the systems that are eligible—a constraint useful only for yielding the relation between chancy facts and deterministic laws.

Can Lewis's account dispense with the "requirement of coherence"? Suppose that although (1) is true, (50%) is not a bad fit to the few ^{346}Un atoms there are. Compare a system containing (50%) to a system that includes both (50%) and (1). The latter is more informative regarding the mosaic of elite-property instantiations, whereas presumably, the former is simpler. Hence, the Best System analysis, without the "requirement of coherence," does not obviously entail that if (50%) is a law, then (1) is not a law.[47] Likewise, compare a system containing (1) to a system that includes both (1) and (50%). They are equally informative regarding the mosaic. However, the latter may be simpler: by including (50%), the system may unify its other statistical claims regarding the lifetimes of various isotopes (and perhaps more) under one simple, broad claim. Without (50%), the unifying claim (and whatever axioms in the deductive system entail it) would have to include an ugly gerrymander ("All nuclei *except* ^{346}Un...". Again, without the "requirement of coherence," the Best System analysis does not obviously entail that if there are ^{346}Un atoms, then (1) is a law only if (50%) is untrue. (We might have expected this result; after all, Lewis felt the need to add the requirement of coherence to his account!)

There is another place from which Lewis might try to derive the relation between chancy facts and deterministic laws. On Lewis's view, the "Principal Principle" (PP) is the key to our concept of chance. It specifies how our opinions about the objective chance of some possible outcome ought to guide our degree of confidence (our "personal probability") that this possible outcome will actually come to pass. Roughly speaking, the PP says that if we believe that there is an objective chance of (say) 20% that a given atom will decay in the next few minutes, then we must (on pain of irrationality) have 20% confidence that it will so decay. More explicitly, the PP says that

for any rational initial subjective probability distribution (cr),

any proposition (A), and

any proposition (E) that expresses a logical possibility,[48] specifies A's present chance to be x, and contains no "inadmissible" information (that is, no information on how any present or future chance process will turn out, apart from the chances of its having various outcomes),

PP $cr(A \mid E) = x$.

Now let E be

a is a ^{346}Un atom that has just now come into existence, and (50%) is true, and it is a law that (1).

Let A be

a decays within 7 μs.

Then PP apparently demands of any rational initial subjective probability distribution cr that $cr(A \mid E) = .5$, since E specifies A's present chance to be 50%. But $cr(A \mid E) = 1$ since E, by virtue of including (1), logically entails A.[49] So PP must be inapplicable to this case—and so E must express a logical impossibility: (50%)'s nonvacuous truth must logically entail that (1) is not a law. Lewis could apparently use PP to explain the relation between chancy facts and deterministic laws.

Not so fast. We do not need E to be logically impossible in order to render PP inapplicable to this case. PP is already inapplicable if E is inadmissible. And E *is* inadmissible: by including (1), E contains information (going beyond the chances of various possible outcomes) regarding how some present or future chance process will turn out.

It might be objected that PP is supposed to be the principle by which our beliefs about chances should ordinarily guide our degrees of confidence. Background beliefs about deterministic laws must ordinarily count as admissible—else we could not ordinarily employ PP, since we have a good many such beliefs. The only "inadmissible" information is supposed to be a certain kind that we do *not* ordinarily have—the sort that would be gleaned from time-traveling to the future or gazing into crystal balls that somehow reveal the future outcomes of chance events. We ordinarily have information about deterministic laws.

However, I reply, those deterministic laws do *not* include information about how future chance events will turn out—precisely because of the relation between chancy facts and deterministic laws that we are trying to explain. We are implicitly *presupposing* that relation in deeming deterministic laws admissible.[50] Insofar as PP counts information regarding deterministic laws as admissible, PP cannot be used to explain the relation between chancy facts and deterministic laws. Deterministic laws are admissible not because any information whatever about deterministic laws automatically counts as admissible, but because deterministic laws cannot contain information about the outcomes of chance processes.

Here is another proposal for using PP to derive the relation between chancy facts and deterministic laws. Consider

(100%) Every ^{346}Un atom, at its creation, has a 100% chance of decaying within 7μs.

Suppose for *reductio* that (50%)'s nonvacuous truth is logically consistent with (100%)'s truth. Let E be

a is a ^{346}Un atom that has just now come into existence, (50%) is true, and (100%) is true.

Let A be

a decays within 7 μs.

Then PP demands of any rational initial subjective probability distribution that $cr(A|E) = .5$ and $cr(A|E) = 1$. But then cr is not a rational subjective probability distribution—a contradiction. So, by *reductio*, (50%)'s nonvacuous truth and (100%)'s truth are logically incompatible.[51]

Now add that (1)'s lawhood logically entails (100%)'s truth. It follows that (50%)'s nonvacuous truth logically precludes (1)'s lawhood.

But this explanation of the relation we have identified between chancy facts and deterministic laws presupposes another relation between chancy facts and deterministic laws: that (1)'s lawhood logically entails (100%)'s truth. Why does that relation hold? After all, (1)'s *truth* does not logically entail (100%)'s truth (since (1)'s truth is logically consistent with (50%)'s truth). Why, then, does (1)'s *lawhood* logically entail (100%)'s truth?

Since (1) does not logically entail (100%), a deductive system may include (1) without being compelled by logical closure to include (100%). Does Lewis's Best System Account entail that if the *best* deductive system includes (1), then it also includes (100%), and so that (1)'s lawhood entails (100%)'s truth? Compare a system containing (1) to a system that includes both (1) and (100%). The former might well be simpler, and despite being larger, the latter system contains no greater information regarding the mosaic. Lewis stipulates that a system containing exclusively truths like (1), and so ascribing no chances, has perfect fit.[52] Thus, the Best System Account does not obviously entail that if (1) is a law, then (100%) is true. (Again, Lewis apparently agrees, since otherwise, he would have felt no need to add the "requirement of coherence"; even "incoherent" systems could have entered the competition, since they would have lost anyway.[53])

It is tempting simply to eliminate deterministic laws entirely and to treat all laws as statistical—interpreting (1) in the Best System as *meaning* (100%). But this tactic would do violence to Lewis's account. Lewis's approach begins with the mosaic of elite-property instantiations (no chances yet, not even 100%) and then raises some regularities among them (still no chances) to the level of law through a competition for Best System—out of which facts about chances also arise. The Best System is thus a system of facts about the mosaic and truths about chances that arise to improve the system. Deterministic correlations in the system remain distinct from 100% chances.[54]

It seems, then, that Lewis's account can explain the relation between chancy facts and deterministic laws only by including a "requirement of coherence" having no other motivation. How bad is that? Arguably, many philosophical accounts include various features having no further motivation than that the rest of the account needs them in order to yield some uncontroversially correct answer. But ideally, each of an account's basic features performs many tasks, not just one, and is difficult to alter or to eliminate without violating the account's overall spirit. In contrast, the requirement of coherence seems like an optional, isolated feature that is inserted by hand expressly to make (1)'s lawhood preclude (50%)'s nonvacuous truth.

Let's see if the metaphysical status of chances on Lewis's picture would explain the relation between chancy facts and deterministic laws. That a given chance holds at a given moment is nothing more (says Lewis) than a fact about the laws and the history of elite-property instantiations through that moment—namely, that the history and laws entail the chance. That chances are indeed mere creatures of the laws is suggested by our best-confirmed scientific theories involving chances. All of them (whether concerning quantum particles or the progeny of heterozygotes) specify laws by which the history through a given moment fixes the chances at that moment.

On this view, chances are not among the facts "governed" by laws. Rather, they are *constituted* partly by laws; they join the laws in the "government." The relation between chancy facts and deterministic laws involves coordination among parts of the government. In this light, it seems metaphysically innocuous. Let's see if it could be explained by the principle that for every fact of the form $ch_{t1}(A_{t2}) = y$ (that is, y is the chance at time t_1 of elite property A's being instantiated at t_2), there are sufficient laws and elite-property instantiations through t_1 to determine that fact. According to this principle, "the chances are determined" (CD):

CD   If (and only if) $y = ch_{t1}(A_{t2})$, then it is a broadly logical truth that $y = ch_{t1}(A_{t2} \mid \lambda \& h_{t1})$, where λ is logically equivalent to the m's where it is a law that m, and h_{t1} is logically equivalent to the history of elite-property instantiations through t_1.

For example, suppose a given atom at a given moment has a 50% chance of still existing 13.81 seconds later (A_{t2}). This chance is determined: the history to that moment includes that the atom is beryllium-11, and it is a law that for any time t_1, any ^{11}Be atom at t_1 has a 50% chance of still existing at time $(t_1 + 13.81$ seconds). So $y = ch_{t1}(A_{t2} \mid \lambda \, \& \, h_{t1})$ in this case amounts to $.5 = ch_{t1}(A_{t2} \mid [.5 = ch_{t1}(A_{t2})])$, which is a logical truth.[55]

Let's see how CD would account for the relation between chancy facts and deterministic laws. If $ch_{t1}(A_{t2}) \neq 0$, then by CD, it is a broadly logical truth that $ch_{t1}(A_{t2} \mid \lambda \, \& \, h_{t1}) \neq 0$. But if the laws logically entail $\sim A_{t2}$, then $ch_{t1}(A_{t2} \mid \lambda \, \& \, h_{t1})$ amounts to $ch_{t1}(A_{t2} \mid \sim A_{t2})$. If "ch" is (as a matter of broadly logical truth) a probability function, then $ch_{t1}(A_{t2} \mid \sim A_{t2}) = 0$, so we have a contradiction. Thus, it follows from CD (and that "ch" must be a probability function) that if $ch_{t1}(A_{t2}) \neq 0$, then the laws cannot logically entail $\sim A_{t2}$—the relation between chancy facts and deterministic laws.

But is CD metaphysically necessary?[56] Our best-confirmed scientific theories involving chances accord with CD. But that does not show CD to be *metaphysically* compulsory.

The most obvious way for CD to be metaphysically necessary is for the fact that a given chance holds to be nothing more than the fact that the laws and history entail it. However, in scientific practice, chancy facts function in many respects unlike parts of the government. Rather, they behave just as any law-governed facts do. A chancy fact may serve as an initial condition in a scientific explanation of an elite-property instantiation and may in turn be explained by laws and initial conditions consisting of elite-property instantiations (as when an atom's half-life or a die's chance of landing on six is explained). A regularity involving chancy facts may be accidental or a matter of law. (For example, it is accidental that every atom in a given vial at a given time has a 50% chance of decaying within 7 μs.) Nothing here suggests that facts about chances are ontologically not on a par with facts concerning the mosaic of elite-property instantiations.

It is more difficult to understand how facts about chances manage to function in scientific practice just like facts about the mosaic if facts about chances are reduced to certain facts involving the laws. For example, suppose that a given atom now has a 50% chance of undergoing radioactive decay within the next 13.81 seconds and we want to know why. An innocuous-looking explanation is that the atom is ^{11}Be and it is a law that ^{11}Be's half-life is 13.81 seconds. However, these are the very same facts that *make it true* that the atom's chance of decaying within 13.81 seconds is 50%, if chances are ontologically dependent upon laws and history. For a chance to be scientifically explained by the same facts that make it the chance is for the chance to be scientifically explained by itself. It is widely held that a fact and its explanation must be metaphysically distinct.

With facts ascribing single-case objective chances forming a global mosaic (criss-crossed by regularities, naturally necessary and accidental) just like facts involving elite-property instantiations, one wonders why chances aren't governed by laws just like those other facts—that is, without being reduced to facts involving laws. What makes chances mere creatures of the laws—parts of the machinery of government—rather than ordinary, law-abiding citizens, one among the many varieties of sturdy, ontologically self-sufficient yeomen who are governed by the laws?

Indeed, CD seems like a stubborn holdover from determinism. Once upon a time, determinism gave us our ideal of a well-ordered universe. It was believed that the universe's physical state at any moment is sufficiently rich that it (via the natural laws) determines the universe's physical state at any later moment. We now know that this sort of orderliness is not logically, conceptually, or metaphysically necessary. The universe could be governed fundamentally by irreducibly statistical laws. However, the spirit of determinism dies hard. A vestigial remnant survives in the demand that any facts about the chances at t be determined (via the laws) by the history through t of elite-property instantiations. This demand reins in chances in a manner congenial to the spirit of determinism; though which elite properties are later instantiated may depend on the outcomes of irreducibly chance processes, at least those chances are grounded firmly in the bedrock of elite-property instantiations. But we have made our peace with irreducibly chance processes—with the laws permitting a given history of elite-property instantiations through t to have various possible elite-property sequels after t, sequels that through t had only various chances of coming about. So why should we regard it as metaphysically any more mysterious for the laws to permit a given history of elite-property instantiations through t to have various possible states at t differing *exclusively* in the *chances* they involve—and for those chances to have had until t only various chances of *their* coming about?[57] The notion that any difference between two systems in their $ch_{t1}(A_{t2})$'s must be grounded in a difference in their history through t_1 of elite-property instantiations seems to express the conviction that any difference in outcome (broadly construed) must have an explanation—the same conviction that originally prompted the determinist's ideal of a well-ordered universe.

Although the relation between chancy facts and deterministic laws might seem to be a powerful reason for locating chances within the government, I shall now argue that this relation is best explained by placing chances among the governed—that is, among the subnomic facts.

3.7. How to Account for the Relation

Here is my explanation of the relation between chancy facts and deterministic laws. Suppose that (1) is a law. For the sake of *reductio,* suppose that just like (1), every other m where it is a law that m is logically consistent with (50%)'s nonvacuous truth. Then since Λ is sub-nomically stable, all of its members must be preserved under the counterfactual supposition that (50%) is nonvacuously true. In particular, (1) must be preserved. However, it is not (in some contexts, at least). A ^{346}Un atom with a 50% chance of decaying within $7\,\mu s$ of its creation might have so decayed, but it might just as well not have decayed. It is like a fair coin: had it been tossed, it might have landed heads, but it might just as well have landed tails.[58] So had there existed an atom of ^{346}Un and (50%) been true, then each of the ^{346}Un atoms might have decayed within $7\mu s$, but then again, some might well not have done so. Hence, it is not the case that had (50%) been nonvacuously true, then (1) would still have been true.[59]

Since the conditional entailed by the suppositions we made for the sake of *reductio* turns out to be false, those suppositions cannot hold. Therefore, if (1) is a law, there must be some *other* law m that is logically inconsistent with (50%)'s nonvacuous truth. (Then Λ's sub-nomic stability fails to require that (1) be preserved under the supposition of (50%)'s nonvacuous truth, since that supposition is logically inconsistent with Λ.) We have shown, then, that if (1) is a law, (50%) must be vacuous or false. In other words, if (50%) is nonvacuously true, then (1) cannot be a law. That was the relation between chancy facts and deterministic laws that we were trying to explain.

Of course, the point generalizes beyond (50%)—to

> Every ^{346}Un atom, at its creation, has an n% chance of decaying within 7μs

for any n, and even to

> Some ^{346}Un atom, at its creation, has an n% chance of decaying within 7μs.

For Λ to be sub-nomically stable and contain (1), Λ must contain some other law that is logically inconsistent with the existence of a ^{346}Un atom with an n% chance of decaying within 7μs. That is, if (1) is a law, then (N) is a law:

> (N)   For any n, no ^{346}Un atom at its creation has an n% chance of decaying within 7μs.

Since (N)'s lawhood entails (N)'s truth, and (N)'s truth logically precludes (50%)'s nonvacuous truth, (1)'s lawhood logically precludes (50%)'s nonvacuous truth. That is the relation that we were trying to explain.

This explanation shares the spirit behind Lewis's "requirement of coherence": that laws are not true by chance.[60] Indeed, that (1)'s lawhood entails (N)'s lawhood means that Lewis was correct: the laws never say (1) without also saying that (1) never had any chance of not coming about. But this result is now fully integrated into the rest of the account rather than inserted by hand.[61]

According to the intuitions behind "Nomic Preservation" (NP), from which "sub-nomic stability" sprang in chapter 1, the actual laws of nature would still have been laws had there been nothing but a single lonely electron, though there are other possible lonely-electron worlds where the laws differ (for instance, where the electrostatic force is twice as strong). Thus, which truths are laws and which are accidents fails to supervene on the sub-nomic facts. This picture of the laws as like the rules governing nature's game suggests that the laws are *radically* nonsupervenient on the facts they govern: *any* deductive system drawn from among the governed truths could logically possibly be the *m*'s where it is a law that *m*. In other words, the only constraint that *m*'s lawhood imposes on the governed facts is m. For if "It is a law that m" entailed q, but m did not entail q, then $\sim q$ would be logically consistent with m but rule out m's lawhood—and so a humble member of the governed would be getting above its proper station in unilaterally imposing constraints on the rules.

However, this picture apparently leaves no room for the relation between chancy facts and deterministic laws. I have urged that both (1) and (50%) be counted among the governed. But (50%)'s nonvacuous truth, in precluding (1)'s lawhood without precluding (1)'s truth, is apparently getting above itself by exerting undue influence upon the laws. When (1) and (50%) are nonvacuously true, not every deductive system of truths could logically possibly be the laws. If the governed facts constrain the laws only minimally, but chancy facts constrain the laws considerably, then chances would seem to be located within the government rather than among the governed.

My account reconciles chances being among the governed with the relation between chancy facts and deterministic laws. Admittedly, that relation denies that m is the only constraint on the governed facts that m's lawhood imposes. However, my proposal saves the kernel of the laws' radical nonsupervenience: the laws *as a whole*—p's lawhood plus q's lawhood plus...—impose no constraint on the governed facts besides $p, q, \ldots$. Of course, if one law p is (1), then the other laws $q \ldots$ must include (N), so although (50%)'s nonvacuous truth is logically consistent with (1), it is not logically consistent with all of the claims m that must be laws if (1) is a law.

By this reasoning, a deterministic universe will have many laws like (N)—laws denying that various chancy properties are ever instantiated. Furthermore, there will be many "closure laws" specifying that certain inventories of properties are complete—that each inventory includes every property of a certain sort that is ever instantiated.[62] For example, the laws specifying the characteristic charges, masses, lifetimes, and so forth of the various elementary-particle species require a law that there are no electrically charged leptons besides the ones mentioned in those laws: electrons, muons, and taus. Here is the argument for this law, analogous to the argument I just gave for (N)'s lawhood. By the conservation laws of mass-energy, electric charge, and lepton number, a charged lepton can decay only into another, lighter kind of charged lepton.[63] Thus, the muon and tauon can afford to be unstable, since the electron is a less massive charged lepton for them to decay into. But the electron must be stable (radioactively speaking—not sub-nomically speaking!), since it is the lightest of the three charged leptons. It is a law that all electrons are stable, just as it is a law that all electrons are negatively charged. However, had there been an additional kind of charged lepton besides electrons, muons, and taus, less massive than any of them, then the electron might have been unstable, like the muon and tauon. This conditional's truth (together with the law that the electron is stable) violates NP as well as Λ's sub-nomic stability—unless there is a closure law such as "All charged leptons are muons or electrons or taus." Given that law (or something like it, such as "No charged lepton is lighter than the electron"), the above conditional's antecedent is logically inconsistent with the laws, and so the laws' sub-nomic stability fails to require the laws' preservation under this antecedent.

There must be many laws like "All charged leptons are muons or electrons or taus," precluding unnatural (a.k.a. "alien") kinds of particles, forces, and so forth, just as (1)'s lawhood demands (N)'s. Intuitively, the laws would still have held under any (hypothetical!) twisting of the knobs establishing initial or boundary conditions. (This intuition originally motivated NP.) But to posit a new *kind* of particle—or a new *kind* of property—is not merely to twist the knobs; it is to add a new knob. Twisting the knobs cannot undermine the law that the electron is stable—but adding a knob can.

The existence of closure laws like (N) also seems plausible considering that the laws specify all of the degrees of freedom of the state space. That is, they specify what a system's trajectory through the state space would amount to, which is a prerequisite to their specifying which particular trajectories are possible. Thus, my explanation ties the relation between chancy facts and deterministic laws to other important features of lawhood.

Here is another way to view my explanation of the relation between chancy facts and deterministic laws. If a process is genuinely chancy, then before it yields outcome O, it is not the case that were the process run, it would yield outcome O. For example, if I flip a (genuinely indeterministic) coin and it lands "heads," then before I flipped the coin, it was not the case that were I to flip the coin, it would land "heads." It might have landed "heads," but it might not have.[64] A chancy fact has implications regarding the truth-values (in certain contexts) of subjunctive conditionals. Accordingly, if at time t a given [346]Un atom has a 50% chance of decaying within 7μs, then even if the atom later so decays, it is not the case (at least in certain contexts) that the atom would decay within 7μs were circles round (or—to take another example—were the atom confined to this container for the rest of its life, as it actually is). But this conditional must hold if (1) is a law and the laws form a sub-nomically stable set.

The reason, then, that a chancy fact such as (50%) keeps (1) from being a law, without keeping (1) from being true, is that a chancy fact constrains the *subjunctive* facts and (1)'s lawhood, unlike (1)'s truth, depends upon the subjunctive facts—because of lawhood's relation to sub-nomic stability.

This explanation of the relation between chancy facts and deterministic laws preserves the intuition originally behind the idea that (50%)'s nonvacuous truth logically precludes (1)'s lawhood. If (50%) is nonvacuously true, then a [346]Un atom that lasts for more than 7μs is possible. However, if (1) is a law, then such a [346]Un atom is impossible, by virtue of lawhood's relation to stability and hence to necessity.

Of course, this explanation appeals to the truth-values of certain subjunctive conditionals (that is, to certain "subjunctive facts"). This might seem to get the order of explanation backward: features of the laws should explain features of subjunctive facts rather than the other way around. However, I will devote chapter 4 to arguing that subjunctive facts are the lawmakers.

4

A World of Subjunctives

4.1. What If the Lawmakers Were Subjunctive Facts?

In chapter 1, I argued that

(i) the sub-nomic truths that are laws form a sub-nomically stable set Λ, and

(ii) no sub-nomically stable set contains accidents (except perhaps for the set of all sub-nomic truths).

There is thus a sharp distinction between laws and accidents, and the laws' unique relation to counterfactuals involves no circularity, triviality, or arbitrariness. The association between lawhood and membership in a nonmaximal sub-nomically stable set not only leaves room for multiple strata of natural laws, but also explains why the laws would still have been *laws* under various counterfactual suppositions.

In chapter 2, I suggested that a sub-nomically stable set displays maximal persistence under counterfactual suppositions and so is associated with a variety of necessity. For each species of necessity, the sub-nomic truths so necessary form a sub-nomically stable set, and for each sub-nomically stable set (except the set of all sub-nomic truths, if it be stable), there is a variety of necessity where the sub-nomic truths so necessary are exactly the set's members. Thus, natural necessity and the varieties of necessity possessed by broadly logical truths are species of the same genus. Natural laws are genuinely necessary despite being contingent. The association between nonmaximal sub-nomically stable sets and varieties of necessity not only differentiates genuine from merely conversational necessities, but also explains why the genuine modalities come in a natural ordering.

We now have two associations: first, lawhood is associated with membership in a nonmaximal sub-nomically stable set; and second, such membership is associated with possession of a certain variety of necessity. Why do these associations hold? What is the order of onto-logical priority among lawhood, necessity, and membership in a non-maximal sub-nomically stable set? Is one of these responsible for the others, or do they all arise separately from another, common source?

I suggest that the laws' necessity is what sets them apart from the accidents—what *makes* them laws. (So I intimated in response to the Euthyphro question at the start of chapter 2.) Moreover, I have argued that necessity *consists of* membership in a nonmaximal sub-nomically stable set. Therefore, since *m* is a law in virtue of its necessity, and *m* is necessary in virtue of belonging to a nonmaximal sub-nomically stable set, it follows that *m* is a law in virtue of belonging to a nonmaximal sub-nomically stable set. What, then, is responsible for making true the various subjunctive conditionals that render the set of laws sub-nomically stable? Not the facts about which truths are laws, since they hold in virtue of the set's stability. I propose that with these subjunctive facts, we have reached ontological bedrock. They (along with various sub-nomic facts) are primitive, lying at the bottom of the world. They are the lawmakers.[1]

My proposal reverses the standard picture of laws "supporting" counterfactuals.[2] However, my proposal is perfectly compatible with our ascertaining (or justifying our belief in) a given counterfactual conditional's truth by appealing to our beliefs about the laws. That we figure out which counterfactuals are true by consulting what we already know about the laws (among other things) does not at all sup-port the idea that the truths about the laws are *ontologically* prior to the subjunctive truths.[3]

Of course, my proposal runs counter to the customary view that typical subjunctive facts are not ontologically primitive, but rather are made true by the laws together with various nonsubjunctive ("cate-gorical") facts. For example, the fact that the match would have lit, had it been struck, is made true by various laws about ignition along with the fact that the match is dry, oxygenated, and so forth. But as I men-tioned in chapter 1, the project of analyzing subjunctive facts in terms of nonsubjunctive facts (plus laws) was famously shown by Nelson Goodman to encounter a serious obstacle: the "problem of cotenabil-ity."[4] That the match is actually dry is not directly responsible for mak-ing it true that the match would light, were it struck. That work is done by the fact that the match *would be* dry, *were* it struck. As Goodman emphasized, subjunctive facts seem ineradicable from the truth-conditions of subjunctive conditionals. One way to face up to this result is to locate subjunctive facts at the bottom of the world.[5]

An objection: whether a given counterfactual conditional is true depends upon whether we regard certain facts (for example, that the match is dry) as important, as salient, as facts that contribute mightily toward determining which "possible worlds" where the counterfactual antecedent obtains are "closest" to the actual world. A given fact's salience is not part of the world. Rather, we introduce it as our interests inform various conversational contexts.[6]

This objection amounts to denying that any counterfactual conditionals have objective truth-values. But if there is indeed an association between lawhood and membership in a nonmaximal sub-nomically stable set (regardless of which is ontologically prior), then if counterfactual conditionals are made true partly by us, then presumably so are the facts about what the laws are. *That* I am not prepared to accept![7] Furthermore, to argue that "struck □→ lit" lacks objective truth-value because its truth depends on the truth of "struck □→ dry" plainly begs the question as far as whether counterfactuals lack objective truth-values.

One reason for suspecting that counterfactuals lack objective truth-values is their context sensitivity. Consider an analogous case.[8] Many philosophers deny that there is an objective distinction between the cause of the match's lighting (its being struck) and the standing, background conditions that enabled the struck match to light (such as its being dry and oxygenated). The latter are in the background only insofar as they do not interest us; they remain just as causally relevant as the match's being struck. If the context shifts, then the match's being oxygenated might become *the* cause of its lighting, its being struck demoted to the "causal field." Insofar as which counterfactuals are true seems like which causally relevant factors are *the* causes of the match's lighting, counterfactuals seem to lack objective truth-values.

However, whether the match's being oxygenated is one of *the* causes or instead part of the causally relevant background is fixed entirely by our interests. But which counterfactuals are true is *not* determined wholly by our interests; for example, some counterfactuals are intimately associated with the natural laws. Accordingly, there is more reason to attribute counterfactuals to the world than to attribute the distinction between *the* causes and causal background. Moreover, as I mentioned in chapter 1, the counterfactuals' context sensitivity fails to threaten their objectivity. It merely reflects the fact that the same counterfactual sentence expresses different propositions on different occasions. The subjunctive fact that makes "Had I jumped from the ledge, I would have been injured" true in one conversational context (where the ledge's height and the absence of a net are salient) is distinct from the subjunctive fact that makes this sentence false in another context (where my caution and good sense are salient, so "Had I jumped from the ledge, I would have arranged in advance for a net to catch me" is true). Although a counterfactual conditional's truth-value may depend on the context, its truth *in a given context* is an objective matter.[9]

Sometimes the view that typical subjunctive facts are not ontologically primitive, but rather have categorical facts as their ground, is expressed in slogans such as "truths must have truthmakers"[10] and "truth supervenes on being"[11] where "being" is intended to cover how things actually are, not how things would be, would have been, or might have been. Of course, I can pay lip service to these slogans, agreeing with them as long as "being" is understood to include subjunctive facts.

However, slogans like "All truths are made true by what is" are intended to preclude primitive subjunctive facts. Sympathy for this view is easily evoked. Lewis writes: "Think, for instance, of a metaphysic that reduces the material world to J. S. Mill's 'permanent possibilities of sensation': what, if not the *un*reduced material world beyond the door, could be the truthmaker for a truth about what sensations would have followed the sensation of opening the door? The point seems well taken, at least so far as this example goes. But why stop there?"[12] However, according to the standard view, a typical counterfactual conditional (such as the one about the sensations that would have followed the sensation of opening the door) has its truth-value grounded not solely upon categorical facts, but also upon facts about the laws. Those facts (I have suggested) fail to supervene on the categorical facts. Thus, categorical facts cannot be entirely responsible for the subjunctive facts.

Nevertheless, I shall honor some of the spirit behind these slogans. Part of what animates them is the urge to make do with the most parsimonious class of primitives possible. To include subjunctive facts among the primitives might seem grossly ontologically extravagant. However, if subjunctive truths and facts about the laws fail to supervene on categorical facts, then we have no choice but to admit further primitives. If the laws are distinguished by forming a nonmaximal subnomically stable set, then to ground the lawhood of various sub-nomic facts, we can make do with just the subjunctive facts involving subnomic antecedents and consequents.

What about the reverse option: taking laws (and categorical facts) as primitive and regarding them as together constituting the subjunctive facts?[13] Let's compare this option to my proposal. If the subjunctive facts are primitive and lawhood is analyzed as membership in a nonmaximal sub-nomically stable set, then the special relation between laws and counterfactuals follows automatically. On the other hand, if lawhood is left unanalyzed, then no analysis of what laws are is available to account for that relation. We must insert it by hand—for instance, by stipulating that laws carry a certain special weight in determining which "possible worlds" are "closest" to the actual world. It is more parsimonious to take the subjunctive facts as primitive and lawhood as constituted by membership in a nonmaximal sub-nomically stable set, thereby getting the laws' relation to counterfactuals for free, than to take the facts about the laws *plus* the laws' relation to counterfactuals as primitive.

It might be objected that if subjunctive facts are on the bottom, then we must stipulate that the laws are exactly the members of a nonmaximal sub-nomically stable set, while if laws are on the bottom, then we must stipulate the particular weight they carry in determining which possible worlds are closest to the actual world. So there is no clear advantage to putting subjunctive facts on the bottom. I reply that if subjunctive facts are on the bottom, then there is no need to explain why the laws play a certain role in connection with counterfactuals. The laws simply are whatever facts (if any) play that role (that is, belong to a nonmaximal sub-nomically stable set). Moreover, this particular role is not arbitrary; it is clear why maximal invariance is something special, and hence why we care about identifying nonmaximal subnomically stable sets. In contrast, if the laws are on the bottom, then it remains mysterious what it is about them that gives them any special weight at all (much less the precise weight they do carry) in determining which subjunctive facts hold. There is no reason why the laws play their special role in connection with subjunctive facts. They just do. What are laws that counterfactuals are so mindful of them? (I shall press this point in section 4.4.)

A "primitivist about law" might reply that just as my proposal reduces p's lawhood to a fact about the subjunctive facts, so "law primitivism" reduces a subjunctive fact to a fact about the laws: $q \;\square\!\!\rightarrow p$ is nothing more than the fact that p follows logically from q, the laws, and certain nonsubjunctive facts of a contextually understood sort. Consequently, law primitivism does not have to specify what it is about the laws in virtue of which they deserve to play their characteristic role in connection with counterfactuals. Rather, subjunctive facts are essentially facts that pay special attention to the laws, just as on my proposal, facts about the laws are essentially facts that stand in a certain special relation to the subjunctive facts. Both proposals, then, can explain the laws' special relation to counterfactuals.

However, the apparent parity between the two proposals is an illusion. For one thing, while maximal invariance is clearly special, it is not evident why we should care about facts that essentially pay especial attention to the laws. Primitivists about law cannot appeal to some reductive account of what laws *are* to explain why we should care about facts that privilege laws. Nor can they appeal to an analysis of what lawhood *is* to explain why p acquires a kind of *necessity* by virtue of following from the laws. Suppose we take some other class of facts and identify p as possessing a given variety of "necessity" exactly when p follows from those facts. Why are we more mindful of natural necessity than of this other "variety of necessity"? Why does only the former ground scientific explanations and have a special role in connection with counterfactual conditionals? Primitivists leave themselves few resources for answering such questions.

Furthermore, important kinds of counterfactual conditionals cannot have their truth-values fixed by what follows from the laws, the antecedent, and certain tacitly understood sorts of facts. Consider the subjunctive fact that had there been different fundamental forces, the conservation laws and fundamental dynamical law would still have held (and that it is not the case that their negations might have held). Counterlegals like this one are responsible for the more exalted modal status of the conservation and fundamental dynamical laws as compared to the various grubby force laws. (Recall the pyramidal hierarchy in fig. 1.3.) Consider also the subjunctive fact that had the fundamental dynamical laws been different, certain symmetry principles would still have held (and that it is not the case that their negations might have held). Counterlegals like this one are responsible for the symmetry principles' status as meta-laws, and hence their power to explain why various conservation laws hold (see chapter 3, section 3.5). No nontrivial counterlegal (and, for that matter, no nontrivial counterlogical) can be accounted for by analyzing subjunctives as facts about what follows from the laws, the counterfactual antecedent, and certain tacitly understood sorts of facts.[14]

So far, my defense of locating subjunctive facts on the bottom has been rather (let us say) defensive. In subsequent sections, I shall go on the offensive by describing some of the constructive work that primitive subjunctive facts could do. In section 4.2, I will argue that if subjunctive facts are the lawmakers, then we can explain where the laws' necessity and explanatory power come from. In particular, we can avoid taking the laws' necessity as constituted by facts that are not themselves invariant enough under counterfactual suppositions to bestow necessity upon the laws. In section 4.3, I will elaborate this solution to "the lawmaker's regress" by arguing that the notion of sub-nomic stability can be simplified if subjunctive facts are ontologically on a par with subnomic facts. In section 4.4, I will argue that any view that fails to locate subjunctive facts among the lawmakers will find it difficult to avoid adhocery in its account of the laws' relation to counterfactuals. In sections 4.5 and 4.6, I will argue that instantaneous rates of change (such as velocity and acceleration in classical physics) should be analyzed in terms of ontologically primitive subjunctive facts. With this independent argument that we cannot do without such facts, parsimony favors our putting them to work as lawmakers, too.

Simplicity, parsimony, and an interest in avoiding adhocery are admittedly thin reeds on which to support a glorious metaphysical view. But how else can one argue for taking a certain class of facts as ontologically primitive? Obviously, it is relevant that the candidate primitives cannot be reduced to other facts that we are already prepared to take as ontological bedrock. I have already suggested that subjunctive facts and facts about the laws fail to supervene on the subnomic facts. But such arguments do not suffice to show conclusively that subjunctive facts are primitive; other facts instead might be the lawmakers and be primitive.

All one can hope to show is that one receives a greater return on one's investment by locating certain facts rather than others at the bottom of the world. In sections 4.7 and 4.8, I will consider a final payoff of locating subjunctive facts ontologically alongside sub-nomic facts: the simplified concept of stability at which we thereby arrive explains why the laws must be "complete"—roughly speaking, must have no gaps in their coverage. In section 4.9, I will conclude by considering whether I have cheated by including subjunctive facts among the lawmakers.

4.2. The Lawmaker's Regress

I find primitive necessity more mysterious than primitive subjunctive facts. Why so? A fact seems to be what it is independent of whether it is necessary or not. Its necessity seems to be (speaking loosely) something *added* to the fact. What, then, must be added to a fact to make it necessary? "A primitive necessity" isn't much of an answer. There is no analogous objection to a primitive subjunctive fact: q $\Box\!\!\rightarrow p$. We cannot begin by distinguishing the fact that p from the fact that p would have held had q been the case, and then asking what must be added to the fact that p to make the fact that p would have held had q been the case—for p does not need to hold for q $\Box\!\!\rightarrow p$ to hold. The fact that q $\Box\!\!\rightarrow p$ is not some sort of "strengthening" of the fact that p (or the fact that q). In not being the strengthening of something else, the fact that q $\Box\!\!\rightarrow p$ is more like the fact that p than p's necessity is.

We can sharpen this line of thought to generate a puzzle about necessity—"the lawmaker's regress"—to which I shall devote this section. My approach resolves this puzzle nicely by taking primitive subjunctive facts as responsible for necessity.

The laws' necessity is indispensable to the distinction between laws and accidents. The lawmakers (whatever they are) must be responsible for the laws' necessity. (They do not seem to be needed to make the laws true. That the laws' necessity is a "strengthening" of their truth launched the line of thought above.) But if the lawmakers themselves lack the relevant species of necessity, then it is difficult to see how they can supply the laws with their necessity.

For example, Lewis takes the lawmakers to be the constituents of the Humean mosaic. They lack any sort of necessity—except for what some (but not all) of them ultimately acquire by virtue of following from the laws. Consequently, natural necessity on Lewis's account consists of nothing more than being a logical consequence of the laws. But then why does "natural necessity" amount to a variety of necessity? We cannot just stipulate that being a logical consequence of the laws constitutes being necessary. Natural necessity must *deserve* to count as a species of necessity, as I argued in chapter 2.

Now I shall take this argument one step further: accidents (facts lacking the laws' necessity) cannot make the laws necessary—contrary to Lewis's account. Let me sketch an example. Suppose it is actually a fundamental law that p: all elementary particles of a certain kind ("A-ons") are stable at accelerations below a certain magnitude m but decay instantly if they reach accelerations of m or more. Then the fact F—that the Best System contains exactly p and its colleagues—logically entails that if many A-ons exist, then some A-ons reach accelerations of m or more. That is because if many A-ons exist but none decays or reaches such accelerations, then the Best System contains "All A-ons are stable" instead of p. Addition of p to the Best System would then diminish the system's simplicity far more than boost its strength. Nevertheless, with p as a law, Lewis's account still allows it to be accidental that if many A-ons exist, then some (just before decaying) reach accelerations of m or more. (On Lewis's view, for many A-ons to exist and none to accelerate beyond m is logically consistent with the laws' *truth* though, I have presumed, logically inconsistent with the laws' *lawhood*.) Thus, F can logically entail an accident: that some A-ons reach accelerations of m or more if many A-ons exist. Therefore, F can be accidental.[15] Hence, on Lewis's account, accidents can be responsible for making certain facts naturally necessary.

However, facts without necessity are not equipped to make other facts necessary. Here is a brief argument that a contingent truth cannot be responsible for p's necessity. If F is responsible for p's necessity, then F is responsible for p's holding in all possible worlds. Hence, in any possible world, F is responsible for p's holding. Thus, F holds in any possible world. Therefore, F is necessary. In other words, if F is contingent, then F is not around in some possible world to make p hold there, and so F cannot be responsible in that world for p's holding there (even if p holds there), and so F cannot be responsible for p's necessity.

We can put this argument slightly more formally. To set it up, let's return to the "modality principle" from chapter 2:

(1) p is necessary if and only if p would still have held under any possible circumstance C. (In other words: $\Box p$ if and only if for any C where $\Diamond C$, $C \Box\!\!\rightarrow p$.)

Although any species of modality (logical, natural, and so forth) may be plugged into this principle, the same species of modality must appear throughout a given instantiation. (Obviously, it is not the case that p is *naturally* necessary only if p would still have held under any *logically* possible circumstance.) As we have seen, (1) could be strengthened. For example, suppose it is a matter of natural necessity that no body can travel faster than the speed of light. Then not only had we tried to accelerate a body to superluminal speeds, we would have failed (as (1) says, presuming it possible for us to try to accelerate a body superluminally), but also had we access to 23rd-century technology, then had we tried to accelerate a body superluminally, we would have failed. We could extend (1) to cover that nested counterfactual—that is, to require that under any possibility C', it would still have been the case that had C obtained (for any possibility C), then p would still have held:

(2) $\Box p$ if and only if for any C and C' where $\Diamond C$ and $\Diamond C'$, $C \Box\!\!\rightarrow p$ and $C' \Box\!\!\rightarrow (C \Box\!\!\rightarrow p)$.

This relation between p's necessity and various counterfactuals suggests that if F is p's necessity-maker, then F is responsible for the truth of the various counterfactuals that (2) associates with p's necessity. However, this cannot be quite right, since some of these counterfactuals may be too trivial to require F to secure them. For example, let natural necessity be the species of necessity under consideration and let p be that energy is conserved. Presumably, the truth of "Had a uranium atom in this vial undergone radioactive decay sometime during the past 5 microseconds, then energy would still have been conserved" is secured by whatever fact F makes p naturally necessary. However, another of the counterfactuals $C \Box\!\!\rightarrow p$ that (2) associates with p's necessity is "Had energy been conserved and I worn an orange shirt today, then energy would have been conserved." The truth of this counterfactual conditional is secured not by whatever fact F makes p naturally necessary, but rather by whatever fact makes it logically necessary that energy is conserved if I am wearing an orange shirt today and energy is conserved.

In this case, C logically necessitates p, so F (p's necessity-maker) is not responsible for $C \square\!\!\rightarrow p$. Nevertheless, the nested counterfactuals in (2) include some nontrivial counterfactual conditionals beginning with C for which F *is* responsible, such as "Had energy been conserved and I worn an orange shirt today, then had a uranium atom in this vial undergone radioactive decay sometime during the past 5 microseconds, energy would still have been conserved." Thus, I suggest a slightly refined version of the idea that if F is p's necessity-maker, then F is responsible for all of the counterfactuals that (2) associates with p's necessity:

(3) If F is responsible for $\square p$, then for any C where $\Diamond C$, there is either some truth $C \square\!\!\rightarrow p$, or some truth $C \square\!\!\rightarrow (C' \square\!\!\rightarrow p)$ for some C' where $\Diamond C'$, for which F is responsible.

Furthermore

(4) If F is responsible for $C \square\!\!\rightarrow p$ [or for $C \square\!\!\rightarrow (C' \square\!\!\rightarrow p)$], then $C \square\!\!\rightarrow F$.

For example, if the fact that the match is dry and oxygenated (and that certain laws hold) is responsible for the fact that the match would have lit had it been struck, then it must be the case that the match would still have been dry and oxygenated (and those laws would still have held) had the match been struck. (Otherwise, the fact that the match is actually dry and oxygenated is generally irrelevant to what would have happened, had the match been struck.) We encountered something like this thought in chapter 1: had Earth's axis been nearly aligned to its orbital plane, then although terrestrial seasons would have been quite different, the actual laws of nature would still have been laws—which is *why* terrestrial seasons would have been so different.

Now for the argument that a contingent truth F cannot be responsible for p's necessity. (Again, the basic thought is that if F is responsible for p's necessity, then F is responsible for p's holding in all possible worlds, and so in any possible world, F is responsible for p's holding, and hence F holds in any possible world, and so F is necessary.) If F is responsible for $\Box p$, then by (3), for any C where $\Diamond C$, there is either some truth $C \Box\!\!\rightarrow p$, or some truth $C \Box\!\!\rightarrow (C' \Box\!\!\rightarrow p)$ for some C' where $\Diamond C'$, for which F is responsible. Hence, by (4), $C \Box\!\!\rightarrow F$ for any C where $\Diamond C$. Therefore, by (1), $\Box F$.

Therefore, lawmakers must possess (natural) necessity, else laws could not acquire their necessity from them. But what makes lawmakers necessary? What constitutes their necessity? Any view according to which their necessity is an ontological primitive owes us an account of why their "necessity" deserves to be so called. If a view fails to cash out their necessity in terms of other facts, then it cannot appeal to some analysis of what their "necessity" consists of to explain why it constitutes a species of necessity.

Let's now expose "the lawmaker's regress." If the lawmakers' necessity is constituted by other facts, then those facts are either necessary or not. If they are necessary, then the regress continues as we turn our attention to what makes those facts necessary. On the other hand, if the lawmakers' "necessity" is constituted by facts that are not themselves necessary, then once again, the lawmakers' "necessity" and hence the laws' "necessity" is bogus. In that case, a law has no necessity to bestow upon what it is supposed to explain.

The laws' necessity is crucial to their role in scientific explanations. Long before 1948, when Carl G. Hempel and Paul Oppenheim published their famous deductive-nomological model and covering-law conception of scientific explanation, laws of nature were explicitly recognized as possessing distinctive explanatory power. For example, in his 1841 anatomy textbook, Jacob Henle (the founder of modern histology) wrote, "To explain a physiological fact means in a word to deduce its necessity from the physical and chemical laws of Nature."[16] But for the laws to explain a phenomenon by rendering it unavoidable, inevitable—necessary!—the laws themselves must possess necessity, and so the lawmakers must possess necessity.[17] The lawmaker's regress is thereby launched. If the lawmakers (or the laws themselves) possess their necessity primitively, then we cannot appeal to what that necessity consists of to recognize that it constitutes genuine necessity. On the other hand, if the lawmakers' necessity consists of other facts, then those facts are either necessary or not. But as we have seen, q cannot render p necessary unless q has necessity to lend; otherwise, p's deriving from q gives p only "conditional" necessity. As Simon Blackburn remarks, if p's necessity is constituted by F and "F just cites that something *is* so [but F] does not *have to be* so, then there is strong pressure to feel that the original necessity has not been explained or identified, so much as undermined."[18] Bas Van Fraassen, who is more concerned than Blackburn with scientific explanation, expresses essentially the same problem thus: "To posit a micro-structure exhibiting underlying regularities, is only to posit a new cosmic coincidence. That galvanometers and cloud chambers behave as they do, is still surprising if there are electrons, etc., for it is surprising that there should be such regularity in the behavior of electrons, etc."[19] If the law of energy conservation explains why energy is actually conserved (it *had to be* conserved, so it was)—if it explains why there are no perpetual-motion machines (there *couldn't be* any, that's why there aren't any)—then energy conservation must be a *had-to-be,* and hence so must its lawmakers.[20] (If energy *had* to be conserved merely *given the laws,* then the fact that energy is conserved has mere *conditional* inevitability.) What is the origin of the necessity that percolates upward?[21]

This worry is one motivation for "scientific essentialism," according to which laws possess *metaphysical* necessity. For example, according to scientific essentialism, part of electric charge's essence is that being charged involves having the causal power to exert and to feel forces in accordance with Coulomb's law. The lawmakers, as facts about essences, make their consequences necessary, resolving the lawmaker's regress. However, in trying to pay due respect to the laws' necessity, scientific essentialism ends up assigning laws *too much* necessity. As I mentioned in chapter 2, scientific essentialism fails to place the laws *between* the broadly logical necessities and the accidents. (I will argue later that scientific essentialism also fails to give a satisfying explanation of the laws' relation to counterfactuals.) But essentialism does at least address the lawmaker's regress. I shall now offer a resolution of the lawmaker's regress that not only avoids positing primitive necessity, but also avoids assigning the laws either too much necessity or too little.

On my view, if it is a law that p, then p's necessity consists of p's membership in a nonmaximal sub-nomically stable set Λ. Therefore, p is necessary in virtue of various subjunctive facts $q \; \square\!\!\rightarrow p$, $r \; \square\!\!\rightarrow p$, and so forth. Hence, when p's lawhood explains the fact that p, some of these subjunctive facts are doing the explaining. They explain p by entailing that p is inevitable, is no fluke, is not the result of some accidental circumstance, but rather would still have been the case no matter what—that is, under any possible circumstances. Of course, p would not have been the case under certain logically, metaphysically, mathematically, morally, and conceptually possible (but naturally impossible) circumstances, such as had $\sim p$ been the case. But the circumstances $q, r, \ldots$ that are naturally possible do not form an arbitrary, gerrymandered range. Rather, this range deserves to be called "all *possible* circumstances" because it encompasses exactly those circumstances that are logically consistent with the laws, and the laws' sub-nomic stability (their maximal invariance) invests them with a variety of necessity. Therefore, even though $\sim p$ is logically, metaphysically, mathematically, morally, and conceptually possible, p would still have been the case "no matter what." The subjunctive truths (not merely $q \ \square\!\!\rightarrow p$, $r \ \square\!\!\rightarrow p$, and the others explaining p, but all of the subjunctive facts that make Λ sub-nomically stable) carve out a genuine variety of possibility such that p would still have been the case under any possible circumstance.

How, then, do p's lawmakers manage to constitute p's (natural) necessity? As we have seen, they must themselves be (naturally) necessary. But what constitutes *their* necessity? How does the regress of necessity (the "lawmaker's regress") get resolved?

On the picture I am offering, the subjunctive facts ($q \ \square\!\!\rightarrow p$, $r \ \square\!\!\rightarrow p$, and so forth) that make p a law are *ontologically* primitive (rather than constituted by various categorical facts and laws). But they are not *explanatorily* primitive; they are not unexplained explainers. Like other broadly logically contingent facts, they have scientific explanations; in particular, various facts make them naturally necessary. Just as p is explained by $q \ \square\!\!\rightarrow p$, $r \ \square\!\!\rightarrow p$, and so forth (which ensure that p would still have obtained under all possible circumstances), so $q \ \square\!\!\rightarrow p$ is explained by $r \ \square\!\!\rightarrow (q \ \square\!\!\rightarrow p)$, $s \ \square\!\!\rightarrow (q \ \square\!\!\rightarrow p)$, and so forth (see fig. 4.1). These nested subjunctive truths ensure that ($q \ \square\!\!\rightarrow p$) would still have obtained under all possible circumstances—the very same "all possible circumstances" under which p would still have held! In other words, these nested subjunctive truths give ($q \ \square\!\!\rightarrow p$) the same invariance under counterfactual suppositions as p possesses in connection with its necessity. The subjunctive facts that explain p are thus able to make p inevitable since they, in turn, are inevitable, and the subjunctive facts making *them* inevitable are inevitable, and so forth infinitely—in view of the endless nested counterfactuals involved in Λ's sub-nomic stability. None of these subjunctive facts has mere conditional inevitability. The inevitability of each can be cashed out in terms of other subjunctive facts (all of which are required for Λ to be sub-nomically stable). There is no primitive necessity. Natural necessity is not left unanalyzed; it deserves to count as a species of necessity because of what it is constituted by.

One way to put our original puzzle is that the fundamental laws have no explanations among the laws, making the source of their necessity (and hence their explanatory power) mysterious.[22] I suggest that a law's lawhood is constituted by various subjunctive truths, each of which would still have held no matter what—a fact also among the subjunctive truths that constitute the law's lawhood.

In the next section, I will take this resolution of the lawgiver's regress one step further. Before doing so, let me return briefly to the

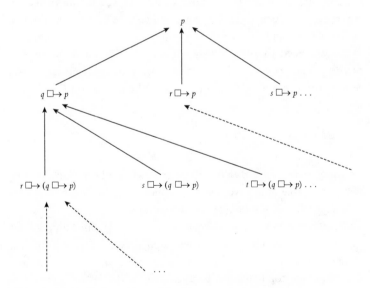

Figure 4.1 If it is a law that p, then various subjunctive facts (only a few of which are shown) explain why p is the case, and for each of these subjunctive facts, various further subjunctive facts (only a few of which are shown) explain why it is the case, and so forth. All of those subjunctive facts help to make it a law that p. (Arrows point from explainer to explained.)

idea that when p's lawhood explains why p is the case, various subjunctive facts $q \; \square\!\!\rightarrow p, r \; \square\!\!\rightarrow p, \ldots$ explain p by making p inevitable: p would still have obtained under any possible circumstances. Let's shift from p to a different explanatory target: What explains why $q \; \square\!\!\rightarrow p$ is the case? Just as p's lawhood explains why p holds, so it should explain why $q \; \square\!\!\rightarrow p$ holds. However, on my proposal, $q \; \square\!\!\rightarrow p$ helps to *constitute* p's lawhood. Thus, for p's lawhood to explain why $q \; \square\!\!\rightarrow p$ holds threatens to involve a fact helping to explain itself. Furthermore, if p's lawhood (that p would still have held no matter what) explains why p holds, then the proper analogy would seem to be for $(q \; \square\!\!\rightarrow p)$'s necessity to explain why $q \; \square\!\!\rightarrow p$ holds. Which is it?

On my view, we do not have to choose. When p's lawhood explains why p holds, the subjunctive facts $q \; \square\!\!\rightarrow p, r \; \square\!\!\rightarrow p, \ldots$ among p's lawmakers do the explaining. By the same token, when p's lawhood explains why $q \; \square\!\!\rightarrow p$ holds, the subjunctive facts $r \; \square\!\!\rightarrow (q \; \square\!\!\rightarrow p)$, $s \; \square\!\!\rightarrow (q \; \square\!\!\rightarrow p), \ldots$ do the explaining. They, too, are among p's lawmakers. They explain why $q \; \square\!\!\rightarrow p$ holds by making $q \; \square\!\!\rightarrow p$ inevitable (by making it the case that $q \; \square\!\!\rightarrow p$ would still have held under any possible circumstances), and $q \; \square\!\!\rightarrow p$ does not thereby help to explain itself.

But I need to say a bit more before I am entitled to identify $(q \; \square\!\!\rightarrow p)$'s persistence under a certain range of counterfactual suppositions with *its* "necessity." That's the goal of the next section.

4.3. Stability

The subjunctive facts that constitute p's lawhood by making Λ sub-nomically stable do not yet on my account officially qualify as necessary, since they do not belong to a sub-nomically (or nomically) stable set. They are ineligible for membership in such a set, since they are not sub-nomic (or nomic) facts. So although they have enough modal force to constitute p's necessity, satisfying the demand that no accident be responsible for making p necessary (since they exhibit the same invariance under counterfactual suppositions as p does in connection with its necessity), we cannot yet say that their *necessity* bestows necessity upon p.

We could deal with this omission simply by stipulating that for any subjunctive conditional ω (that can be constructed exclusively out of sub-nomic claims and logical connectives, "$\Diamond\!\!\rightarrow$", and "$\Box\!\!\rightarrow$"), ω's truth possesses the same species of necessity as the members of some non-maximal sub-nomically stable set Γ exactly when Γ's sub-nomic stability entails that (in every context) ω would still have held under every sub-nomic counterfactual supposition that is logically consistent with Γ—that is, under every possible circumstance.

But there is another, more attractive option. I have proposed that certain subjunctive facts lie alongside sub-nomic facts at the bottom of the world. So we should consider the counterfactual invariance of sets consisting of *both* sub-nomic truths and subjunctive truths (expressed by subjunctive conditionals made exclusively of sub-nomic claims and logical connectives, "$\Diamond\!\!\rightarrow$"'s, and "$\Box\!\!\rightarrow$"'s).[23] Likewise, to treat these subjunctives on a par with sub-nomics, we should consider a set's invariance not merely under suppositions positing that certain sub-nomic claims are true, but also under suppositions positing that certain subjunctive conditionals are true. Doing so turns out to simplify our notion of "stability." Let's see how.

In chapter 1, we concluded that certain counterfactuals having counterfactuals as their consequents (such as $q \Box\!\!\rightarrow (r \Box\!\!\rightarrow m)$) must be true in order for m to behave like a law in connection with counterfactuals. Accordingly, I built those nested counterfactuals into the definition of "sub-nomically stability." For analogous reasons, I might have required that certain counterfactuals having counterfactuals as their *antecedents* (such as $(p \Box\!\!\rightarrow q) \Box\!\!\rightarrow m$) be true in order for m to behave like a law in connection with counterfactuals. Suppose, for example, that I am holding a well-made, oxygenated, dry match under ordinary conditions. The match would have lit, had it been struck. But had it been the case that the match would *not* have lit, had it been struck, then … what? The match would (have to) have been wet or incorrectly made or starved of oxygen. The actual laws of nature would still have been laws, had it been the case that the match would not have lit had it been struck. (Indeed, the *reason* why the match would have to have been wet or incorrectly made or starved of oxygen is precisely *because* the actual laws of nature would still have been laws.) Thus, for Λ's stability to capture the laws' characteristic resilience under counterfactual suppositions, Λ's stability needs to require Λ's invariance under certain suppositions positing the truth of various counterfactual conditionals.

Conditionals with counterfactual conditionals in their antecedents might initially appear exotic, but actually they are not. Consider counterfactuals with antecedents involving dispositions, such as "Had the box contained a fragile vase, then I would have taken great care not to drop it, since the vase might well have broken had the box been dropped." Even if the ascription of a disposition is not logically equivalent to some counterfactual conditional, dispositions seem to involve "threats and promises"[24] and hence have a counterfactual flavor. A counterfactual antecedent positing the truth of a counterfactual conditional is no more exotic than "Had the box contained a fragile vase." Likewise: "Had the boiling point of the liquid in our test tube been under 300K (under standard pressure), then it would have already boiled by now." This counterfactual conditional is perfectly innocuous, yet its antecedent is "Had it been the case that the liquid in our test tube would boil were it heated to 300K under any standard conditions."[25]

Of course, Λ would *not* still have held had it been the case that had I sneezed a moment ago, a body would then have been accelerated from rest to beyond the speed of light! But of course, the counterfactual conditional in this conditional's antecedent is logically inconsistent with one of the counterfactual conditionals required by Λ's sub-nomic stability. Thus, in order for Λ to be distinguished from sets containing accidents by its invariance under various counterfactual suppositions, including some positing various subjunctive conditionals to be true, Λ had better not have to be preserved under a supposition positing the truth of a subjunctive conditional that is logically inconsistent with one of the subjunctive conditionals required by Λ's sub-nomic stability.

Our aim is to refine the concept of stability so that it not only captures the laws' characteristic resilience under counterfactual suppositions, but also treats subjunctive claims (expressed entirely in terms of sub-nomic claims and logical connectives, "$\Diamond\!\!\rightarrow$", and "$\Box\!\!\rightarrow$") on a par with sub-nomic claims. (I'll use lower-case Greek letters ω, ϕ, and so forth for both such subjunctive claims and sub-nomic claims.) That is, we want to make such subjunctive truths eligible to be included in a "Stable" set (with a capital "S," to distinguish it from our prior notions); any included in a Stable set would then express subjunctive facts possessing the same necessity as the set's sub-nomic truths.

Letting Γ be a nonempty set of claims ω containing every logical consequence ζ of its members, we can define "Stability" along roughly the following lines:

Γ is *Stable* exactly when there are some ϕ where $\Gamma \cup \{\phi\}$ is logically consistent and for any such ϕ and any member ω of Γ, $\sim(\phi$ $\diamond\!\!\!\rightarrow \sim\omega)$ holds (in any conversational context).

In short, Γ is Stable exactly when its members are preserved together under every subjunctive or counterfactual supposition ϕ under which they *could* logically possibly be preserved together.[26]

Our reasons in chapter 1 for holding that no nonmaximal set containing accidents achieves sub-nomic stability carry over as reasons for holding that no nonmaximal set containing accidents possesses Stability. Accordingly, I suggest that it is a law that m if and only if m belongs to a Stable set that does not include all of the sub-nomic truths. Because a Stable set is as resilient under counterfactual suppositions ϕ as it could logically possibly be, its members possess a variety of necessity (as long as the set is nonmaximal in that it does not contain every sub-nomic truth). Thus, if a nonmaximal Stable set Λ° includes Λ's members and all of the subjunctive truths responsible for Λ's sub-nomic stability, then those subjunctive truths possess the same flavor of necessity as Λ's members.

What sort of set that includes Λ's members might possess Stability? Suppose it is a law that m. Let ϕ be a subjunctive claim: that it is not the case that m would have held, had there obtained some arbitrary hypothetical accident—say, had Jones missed his bus to work this morning. Of course, ϕ is actually false; since m is a law, m would still have held had the accidents been different. Accordingly, had ϕ held, then either m would not still have been a law or the laws would have been different in some other respect (so as to render ϕ's antecedent naturally impossible). But had there been no law to make m's truth mandatory, then m might not still have been true. (And, of course, had m not been true, then m would not have been a law.) Therefore, had ϕ held, then the actual laws might not still have been true. So for a set Γ containing Λ's members to be Stable, Γ must also contain the counterfactual $\sim\phi$ ("Had Jones missed his bus to work this morning, then m would have held"), the truth of which helped to make Λ sub-nomically stable. Since Γ includes $\sim\phi$, $\Gamma \cup \{\phi\}$ is not logically consistent, and therefore the laws do not have to be preserved under ϕ for Γ to be Stable.

Thus, for Γ to be Stable, Γ must contain all of the counterfactual conditionals $p \,\square\!\!\rightarrow m$ for every p where $\Gamma \cup \{p\}$ is logically consistent and every m in Λ. This argument concerning Γ's member m can now be applied to Γ's member ($p \,\square\!\!\rightarrow m$). Let ϕ now be that it is not the case that ($p \,\square\!\!\rightarrow m$) would still have held, had there obtained some arbitrary hypothetical accident—say, had Smith worn an orange shirt this morning. Repeating the earlier argument, we find that had ϕ, then the actual laws might not still have been true. So to be Stable, Γ must also contain the counterfactual ~ϕ ("Had Smith worn an orange shirt this morning, then ($p \,\square\!\!\rightarrow m$) would have held"), the truth of which helped to make Λ sub-nomically stable. Since Γ includes ~ϕ, $\Gamma \cup \{\phi\}$ is not logically consistent, and therefore the laws do not have to be preserved under ϕ for Γ to be Stable. Thus, for Γ to be Stable, Γ must contain all of the counterfactual conditionals $q \,\square\!\!\rightarrow (p \,\square\!\!\rightarrow m)$ for every p where $\Gamma \cup \{p\}$ is logically consistent, every q where $\Gamma \cup \{q\}$ is logically consistent, and every m in Λ.

And so on, for further nested subjunctives. Thus, if we are building a Stable set Γ by starting with Λ's members, we must add all of the conditionals in the cascade

$$p \,\square\!\!\rightarrow m,$$
$$q \,\square\!\!\rightarrow (p \,\square\!\!\rightarrow m),$$
$$r \,\square\!\!\rightarrow (q \,\square\!\!\rightarrow (p \,\square\!\!\rightarrow m)),\ldots$$

for every member m of Λ and every $p, q, r, \ldots$ where $\Gamma \cup \{p\}$ is logically consistent, $\Gamma \cup \{q\}$ is logically consistent, $\Gamma \cup \{r\}$ is logically consistent, and so forth. Moreover, having added ($p \,\square\!\!\rightarrow m$) to Γ, we must also add ~ ($p \,\Diamond\!\!\rightarrow \sim\!m$), since had the latter failed to hold, the former might have failed to hold—and similarly for the other members of the cascade. Thus we have added to Γ exactly the counterfactual conditionals required for Λ's sub-nomic stability. Those subjunctive truths will then possess the same variety of necessity as Λ's members.

Every subjunctive conditional that we have so far added to Γ has had a sub-nomic claim as its antecedent. What about conditionals of the sort I introduced near the start of this section: those having subjunctive conditionals in their antecedents? We also need to include some of them in Γ, such as "Had it been the case that a match at spatiotemporal location L would not have lit, had it been struck, then Coulomb's law would still have held" ($\phi \: \square \rightarrow m$). This conditional is true; had it been the case that the match would not have lit, had it been struck, then the match would have been wet or deoxygenated or…, but the natural laws would still have held. Had $\phi \: \square \rightarrow m$ been false, then the laws would have been different; Coulomb's law might not still have held. So to be Stable, Γ must include $\phi \: \square \rightarrow m$ so that Γ's Stability does not require its invariance under the supposition that ($\phi \: \square \rightarrow m$) is false.

What, then, would constitute a Stable set Λ° that includes all of the laws m? It must include not only the above cascade of conditionals that must be true for Λ to be sub-nomically stable, but also every $\phi \: \square \rightarrow m$, $\phi' \: \square \rightarrow (\phi \: \square \rightarrow m)$, and so forth, where ϕ, ϕ', and so forth are each logically consistent with the cascade.

Admittedly, this Stable set is more difficult to inventory than the nonmaximal sub-nomically stable set Λ. On the other hand, by considering sets that include subjunctive facts right alongside sub-nomic facts, we make "Stability" simpler to define than "sub-nomic stability." Since a Stable set contains the cascade $p \: \square \rightarrow m, q \: \square \rightarrow (p \: \square \rightarrow m), \dots$, our definition of "Stability" (unlike our definition of "sub-nomic stability") has no need to expressly require that a Stable set's members be preserved under *nested* counterfactual suppositions. That the member ($p \: \square \rightarrow m$) is preserved under q means that $q \: \square \rightarrow (p \: \square \rightarrow m)$ holds and so automatically ensures that the member m is preserved under this nested pair of counterfactual suppositions.

Thus, the lawmakers (as members of a Stable set Λ°) are necessary, resolving the "lawmaker's regress" I introduced in the previous section. Each lawmaker ω derives its necessity from various other lawmakers that together express the fact that ω would still have held no matter what—that is, under every possible circumstance. Those other lawmakers are themselves necessary (and so are fit to render ω necessary) by virtue of yet other lawmakers doing the same for them (see fig. 4.1). Each of the subjunctive facts that helps to constitute m's necessity is itself necessary, its necessity constituted by other subjunctive facts that help to constitute m's necessity. All of these lawmakers belong to Λ°.

In this way, the lawmakers (consisting of a heap of subjunctive facts) depend on no outside facts to constitute their necessity. They are self-sufficient as far as their necessity is concerned. This is part of what makes the laws' necessity appear so puzzling in the first place. Where does it come from? If we locate its source in facts that are not necessary (as Lewis's account does), then the laws' necessity evanesces. If we locate its source in primitive necessities (as Armstrong's account does, with relations of "nomic necessitation" among universals), then it remains unclear what *makes them* necessities. Scientific essentialism locates the source of the laws' necessity in facts that we already recognize as genuine necessities—but that is because they are metaphysical necessities, and hence necessities of an altogether different species from laws. The lawmaker's regress is hard to resolve because there seem to be no other facts supplying the fundamental laws of nature with their necessity. Now we can see why: the lawmakers as a whole are self-contained. Since every lawmaker is necessary exclusively in virtue of other lawmakers, the laws are able to render certain regularities necessary without having to derive their necessity from anywhere else.

4.4. Avoiding Adhocery

In chapter 1, I promised to offer you some recipes for generating worries about many proposed accounts of natural law. In chapter 2, I presented a recipe directed against any account that answers the Euthyphro question ("Are the laws necessary by virtue of being laws, or are they laws by virtue of being necessary?") by deeming lawhood to be metaphysically prior to natural necessity. To follow that recipe, we ask how the alleged lawmakers manage to make the laws (or their consequences) *necessary*. Now I would like to give you another recipe for causing trouble. It can be directed against any account according to which the lawmakers do not explicitly involve subjunctive facts.

In chapter 1, we found a way to capture the laws' special relation to counterfactuals: in terms of sub-nomic stability. Any account of laws should explain why they form a sub-nomically stable set. But any account in which the lawmakers do not explicitly include subjunctive facts will have difficulty in meeting this demand.

Such an account may succeed tolerably well at explaining why the laws tend to persist under counterfactual perturbations. But *that is not enough*. The account should explain why the laws have *precisely* the perseverance characteristic of laws, no more and no less. In other words, the account should entail not merely that the laws carry some vague extra influence in determining which possible worlds are "closest." The account should entail the *particular weight* they carry: the specific range of counterfactual suppositions (in various contexts) under which the laws are characteristically preserved.

One way, then, that an account can fail to explain the laws' relation to counterfactuals is for it to fail to reach its specific explanatory target. To avoid such failure, the account may resort to adhocery: the laws' precise range of counterfactual invariance may be artificially built into it. This tactic should not be tolerated. If an account proposes that laws essentially involve Best Systems or essences or relations among universals or whatever, then those metaphysical ingredients should neatly entail that the laws form a sub-nomically stable set. If there is no independent characterization of (say) the nomic-necessitation relation among universals from which it follows that all actual nomic-necessitation relationships would still have held under any supposition *p* that is logically consistent with every *m* where it is a law that *m*, then this explanatory deficiency cannot be addressed by having the account stipulate expressly that the actual nomic-necessitation relationships would still have held under all of those suppositions. The laws' special relation to counterfactuals should not have to be put in by hand. It should fall nicely out of the account.

Any account that does not take subjunctive facts themselves to be the lawmakers faces the challenge of getting the right answer (that the laws form a sub-nomically stable set) without adhocery. Obviously, if the laws just *are* the truths forming a nonmaximal sub-nomically stable set, then no independent metaphysical analysis of lawhood must be forced artificially to yield the laws' stability. That represents a considerable advantage for my proposal, I think.[27]

Scientific essentialism fails to meet this challenge. According to essentialism, laws specify the properties essentially bound up with membership in various natural kinds and the causal powers essential to possessing various properties. For example, essentialism says that to be an electron is to possess certain quantities of electric charge, mass, and so forth, and that it is essential to electric charge that like charges at rest exert mutually repulsive forces declining with the square of their separation. How does essentialism account for the laws' stability? As I explained in chapter 2, section 2.12, essentialism says that had I worn an orange shirt, then the natural kinds would still have been the actual natural kinds and therefore the laws would have been no different. But why would the natural kinds have been the same? To reply that the same natural kind of world (that is, a world with the same natural kinds of objects, properties, and processes) would have existed, had I worn an orange shirt, is to leave us with the same sort of question as before: why would the same natural kind of world have existed, had I worn an orange shirt?

Essentialists sometimes reply that *obviously,* the closest possible world where I wear an orange shirt is of the same natural kind as the actual world.[28] The actual world is automatically closer to any possible world of the same natural kind than to any possible world of a different natural kind. But this reply begs the question. What makes a world's "essence" (the natural kinds there, the laws there) more influential than other considerations in determining which other worlds are "closest" to it? To dress up the laws as part of the world's essence does not tell us why the laws would still have held, had I worn an orange shirt, unless we already know why a world with the same essence would have existed, had I worn an orange shirt. Rather than stipulating that it would, we should answer the original question: why would the laws still have held, had I worn an orange shirt?

To answer that question properly, essentialism must explain the laws' *particular* relation to counterfactuals. That the laws specify the properties essentially bound up with membership in various natural kinds presumably gives them *some* special influence in determining which possible worlds where I wear an orange shirt are closest. But even so, I do not see how essentialism can explain the laws' specific relation to counterfactuals without first undergoing considerable ad hoc tinkering. Many different principles assign the laws special weight in determining which possible worlds where I wear an orange shirt are closest. How can the laws' status as essential to various natural kinds determine which of these principles are true? (See fig. 4.2.)

Each of the principles that can be generated from figure 4.2 (and many others) assigns laws some special weight in fixing a given possible

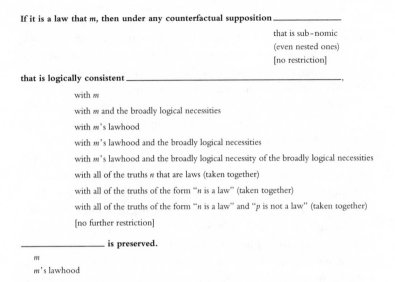

If it is a law that *m*, then under any counterfactual supposition _____

> that is sub-nomic
> (even nested ones)
> [no restriction]

that is logically consistent _____ ,

> with *m*
> with *m* and the broadly logical necessities
> with *m*'s lawhood
> with *m*'s lawhood and the broadly logical necessities
> with *m*'s lawhood and the broadly logical necessity of the broadly logical necessities
> with all of the truths *n* that are laws (taken together)
> with all of the truths of the form "*n* is a law" (taken together)
> with all of the truths of the form "*n* is a law" and "*p* is not a law" (taken together)
> [no further restriction]

_____ **is preserved.**

> *m*
> *m*'s lawhood

Figure 4.2   Fill in the blanks with any of the candidates displayed below them to generate various ways in which a law could carry special weight for counterfactuals.

world's "closeness" to the actual world.[29] Thus, each of them could happily be accommodated by the view that laws differ from accidents by expressing the natural-kind structure. But this flexibility is a sure sign of explanatory weakness, not strength. If essentialism could equally well explain the laws' distinctive relation to counterfactuals *whichever* principle above turns out to capture that relation, then essentialism cannot explain why the laws stand in one of these relations rather than another. It is difficult to see how essentialism's general metaphysical picture could discriminate among the options finely enough to explain why the truths secured by essences behave in accordance with one of these principles rather than another.

Here is another way to display essentialism's explanatory impotence. As we saw in chapter 1, various accidents display *some* degree of resilience under counterfactual perturbations. For example, the relation between my car's acceleration on a dry, flat road and the distance of my car's accelerator pedal from the floor, despite being an accident, would still have held had I depressed the accelerator pedal a bit farther toward the floor or had I worn an orange shirt. By the same token, it is not the case that a law is preserved under *every* supposition. What feature of a law's behavior in counterfactuals is symptomatic of its reflecting the natural-kind structure considering that an accident can be plenty resilient without reflecting the natural-kind structure?

Of course, essentialism could simply *stipulate* that essences are the sorts of things that, by making *m* true, compel *m* to belong to a nonmaximal sub-nomically stable set. But this approach brings to mind the famous quip from Bertrand Russell: "The method of 'postulating' what we want has many advantages; they are the same as the advantages of theft over honest toil."[30] An essentialism burdened by this sort of ad hoc fine-tuning has merely had the right answer inserted into it by hand.

Essentialism might be understood as entailing that possible worlds with greater overlap of natural kinds are "closer" than those with less overlap of natural kinds—in other words, that the closest worlds where *p* obtains are worlds with as many natural kinds as possible in common with the actual world.[31] But if this principle follows from essentialism, then essentialism is mistaken! Counterfactuals like the following are commonly asserted in physics:

> Had the electron's charge been 5% greater, then the energy levels of the electrons within a silicon atom would have been 8% closer together, and so an electron would have needed to acquire less energy to ascend to a higher-energy orbital.

Obviously, in a world where the electron's charge is 5% greater, the actual law specifying the electron's charge cannot be a law.[32] But all of the other actual fundamental physical laws could still be laws, and so the "maximize overlap of natural kinds" principle apparently demands that they all be laws in the closest such world. However, they are not all laws in that world. In particular, the actual law specifying the *proton's* electric charge is not a law in the closest possible world where the electron's charge is 5% greater, since the above counterfactual's truth requires that the silicon atom remain electrically neutral in that world, and so that the proton's charge keep pace there with the electron's. Therefore, the closest possible world where the electron's charge is 5% greater has less overlap of fundamental natural kinds than a more distant possible world where the electron's charge is 5% greater.

Can essentialism be disposed of so easily—by showing that it entails a false counterfactual? Of course not. Essentialism can easily be reconciled with this counterfactual's falsehood: by denying that essentialism entails the "maximize overlap of natural kinds" principle. However, the ease with which essentialism can be decoupled from this principle dramatizes how little essentialism does to explain the laws' relation to counterfactuals. Essentialism could be coupled with nearly any relation of laws to counterfactuals that assigns laws some special weight in fixing a given possible world's "closeness" to the actual world. That is because essentialism entails none of these relations.

One strategy for making essentialism account for the laws' stability is to argue that if the laws reflect the essences of the natural kinds, then the laws are metaphysically necessary, and (by chapter 2's argument) for any variety of necessity, the sub-nomic truths so necessary form a sub-nomically stable set. However, this explanatory route has two main drawbacks. First, as we saw in chapter 2, essentialism thereby places all of the laws on a modal par with one another (and with whatever metaphysical necessities there may be). It flattens the pyramid of sub-nomically stable sets, incorrectly treating all varieties of natural necessity as equally strong. Second, in this proposal, all of the explanatory work seems to be done by the laws' necessity. Essences make no contact with the counterfactuals being explained except through the laws' necessity. What, then, is lost by leaving essences out of the proposal altogether? Of course, the essences are supposed to *make* the laws necessary. But even if essences are dropped from the proposal, subjunctive truths remain, and it would be more parsimonious for *them* to make the laws necessary. If sub-nomic stability cashes out the laws' necessity, then essences are out of a job.

Multiple strata of first-order laws and meta-laws (which I discussed in chapter 3) could be accommodated by Lewis's Best System Account. Suppose that a deductive system containing exactly Λ's members is tied for Best with a system that omits the force laws but includes the fundamental dynamical law, various conservation laws, and the law of the composition of forces, among others. Then Lewis's account might be amended to entail that there are two strata of natural laws. Likewise, meta-laws (such as symmetry principles) could form the Best System of truths *about the first-order laws* (that is, the Best System of truths about the Best System of truths about the Humean mosaic).

Essentialism, on the other hand, may find it more difficult to accommodate these features without adhocery. Essentialism takes counterfactuals such as "Had I worn an orange shirt, then gravity would still have declined with the square of the distance" to be grounded in essences (in this case, gravity's). The actual kinds of forces are supposed to be fixed by the *world's* essence, making it true that the same kinds of forces would still have existed had I worn an orange shirt. But even the world's essence would seem unable to make it true that a given symmetry (or the fundamental dynamical law or a conservation law) would still have held *had there been different kinds of forces*. According to essentialism, a possible world with different kinds of forces is *not* of the same natural kind as the actual world. So even if essentialism could avoid adhocery in embracing the principle that a counterfactual conditional holds exactly when its consequent holds "in a world of the same natural kind as ours in which the antecedent condition is satisfied, other things being as near as possible to the way they actually are,"[33] this principle cannot account for meta-laws and multiple strata of first-order laws.

4.5. Instantaneous Rates of Change and the Causal Explanation Problem

I shall now pursue a different approach to defending my picture of subjunctive facts as primitive. Many philosophical projects aim to show that a certain class of allegedly problematic facts is grounded exclusively in a certain class of allegedly unproblematic facts. A classic way to thwart such a project is to show the "unproblematic" class to be irremediably contaminated by facts of the "problematic" kind. (This strategy was famously deployed in critiques of phenomenalism and operationism, for instance.) In this section and the next, I shall deploy this strategy to argue that subjunctive facts cannot be grounded exclusively in an "unproblematic" base that includes facts about the instantaneous rates of change of various quantities— paradigmatically, facts about instantaneous velocity and acceleration in classical physics. I will argue that unless these facts are themselves constituted by irreducibly subjunctive facts, they cannot play the causal and explanatory roles traditionally ascribed to them in classical physics.

Of course, I do not maintain that classical physics, outfitted with absolute space and time, accurately describes the actual fundamental physical laws.[34] However, I believe that classical velocity and acceleration are exemplars of instantaneous rates of change. Whatever treatment they should receive probably carries over, mutatis mutandis, to the instantaneous rates of change of many continuously changing quantities: the rate at which your bathtub's water level is rising, the rate at which the temperature of the water in your teakettle is rising, the rate at which a wire's electrical resistance is increasing, the rate at which an ecosystem is fixing energy in photosynthesis, the rate at which your impatience is rising, the rate at which an inflating balloon's volume is increasing—perhaps even the rate at which the national debt is increasing. These instantaneous quantities, though differing in many respects, seem to play roles in causal explanations analogous to those played by instantaneous velocity and acceleration in classical physics as traditionally interpreted.

A body's *average* velocity v over a given temporal interval $[t_1, t_2]$ is the distance it traverses over that interval divided by the size of that interval—in other words, the ratio $[x(t_2) - x(t_1)] / [t_2 - t_1]$, where $x(t)$ is the function (the body's "trajectory") specifying the body's position at time t.[35] But what is its velocity *at a given moment*—its *instantaneous* velocity? On the view found routinely in physics and mathematics textbooks, a body's having a certain instantaneous velocity at time t_0 is reduced to the body's having a trajectory in the neighborhood of t_0 that possesses a certain mathematical feature.[36] In particular, its instantaneous velocity at t_0 is what its average velocity over an interval surrounding t_0 tends toward in the limit as the interval becomes ever shorter. That is,

$$v(t_0) = \lim_{\Delta t \to 0} [x(t_0 + \Delta t) - x(t_0)] / \Delta t$$

where

$\lim_{x \to a} f(x) = L$ if and only if for each positive number ε, there exists a positive number δ such that $|f(x) - L| < \varepsilon$ whenever $0 < |x - a| < \delta$.

In other words, no matter how near to L you dare $f(x)$ to go (as long as you do not require the gap ε between them to become zero), there is some neighborhood (of width 2δ) surrounding a where $f(x)$ meets that challenge throughout that neighborhood.

On this reductive view, there is nothing more to a body's having instantaneous velocity v at t_0 than its trajectory's having v as its "time-derivative" at t_0. Two bodies may be no different intrinsically at t_0 although one is moving at t_0 with nonzero velocity v whereas the other is then at rest. Only the *relations* among one body's positions at various moments in t_0's neighborhood differ from the *relations* among the other body's positions at those moments.

On this view, then, a body's instantaneous velocity brings something about only if a certain relation's holding among points in the body's trajectory brings it about. What does a body's instantaneous velocity bring about? Traditionally, it figures in causal explanations of the body's subsequent trajectory. In classical physics, each body has a definite position, velocity, and acceleration at every instant it exists. In particular, bodies in classical physics move in continuous trajectories; a body does not move from place to place without passing through the intervening space, and a body does not disappear at one moment and reappear only after some finite period of time has elapsed.[37] A body's motion in classical physics is affected by outside influences (force fields, collisions) that cause the body to feel forces. The force that a body feels at a given moment affects the body's instantaneous acceleration a at that same moment according to Newton's second law of motion, $F = ma$, where F is the net force on the body and m is the body's mass. The various instantaneous accelerations that the body undergoes over the course of some period of time cumulatively change the body's instantaneous velocity from what it was at the start of that period. The body's velocity, in turn, causes the body to change its location in a certain way—that is, to follow a certain trajectory.[38]

Thus, in accordance with Newton's second law of motion, a body's trajectory in the interval $<t_0, t_0 + \Delta t]$ can be causally explained by the body's mass, the forces on the body at each moment in the interval $[t_0, t_0 + \Delta t]$, and some initial conditions: the body's position at t_0 and (here comes our concern) the body's velocity at t_0.[39] But as we have just seen, the body's $v(t_0)$ is a cause, under this reductive interpretation, only if the body's trajectory in a neighborhood of t_0 is a cause—and any such

neighborhood includes moments after t_0. Hence, for $v(t_0)$ to be a cause of the body's trajectory in $<t_0, t_0 + \Delta t]$, the body's trajectory in $<t_0, t_0 + \Delta t]$ would have to be a cause of itself. This cannot be. Here we have the germ of a powerful argument against this reductive view: that it cannot account for velocity's causal and explanatory role.[40] I shall call this the "causal explanation problem."[41] Ultimately, it will lead me to conclude that velocity should be understood in terms of primitive subjunctive facts.

I have just suggested that if the body's $v(t_0)$ is a cause of the body's trajectory after t_0, and if the body's $v(t_0)$ is just a relation's holding among points in the body's trajectory at t_0 and neighboring moments, then those points in the body's trajectory must be causes of the body's trajectory after t_0. This suggestion presupposes that a relation's holding is a cause only if the relata are. More explicitly:

> If a cause of e is a relation's holding among a's possessing property A, b's possessing property B, etc., then a's possessing property A is a cause of e, and b's possessing property B is a cause of e, etc. (where a may be identical to b).[42]

This presupposition seems plausible. For example, suppose that object a occupies the left pan of a balance and object b occupies the right. The balance tips toward a because of the relation between a's mass and b's: a's is greater. Our presupposition sensibly demands that a's mass and b's mass each influence the balance. (Newtonian physics says that they exert opposite but unequal forces on it.) Otherwise, it couldn't be that the relation between their masses causes the balance to tip. Likewise, suppose a rock's having greater density than the liquid in which it is immersed causes the rock to sink. Then by our presupposition, the rock's density (as well as the liquid's) is a cause of its sinking. Moreover, the rock's density is nothing more than a relation's holding between its mass and volume. By our presupposition, the rock's mass is a cause of its sinking, and so is its volume.[43]

If $v(t_0)$ is a relation's holding among the points in the body's trajectory at t_0 and neighboring moments, then which neighboring moments are these? In defining $v(t_0)$, we may select an arbitrarily small neighborhood around t_0; for any moment T in $<t_0, t_0 + \Delta t]$, we may select a neighborhood that excludes T. So no particular point of the body's trajectory (other than $x(t_0)$) is indispensable to the body's having $v(t_0)$.

However, this does not make $v(t_0)$ an intrinsic property of the body at t_0. On this reductive view, the body's having $v(t_0)$ depends on the body's locations at instants other than t_0. The view portrays $v(t_0)$ as a relation among the points in the body's trajectory in the interval $[t_0 - \Delta t, t_0 + \Delta t]$ and also as a relation among the points in the body's trajectory in the subinterval $[t_0 - \delta t, t_0 + \delta t]$, where $\Delta t > \delta t > 0$. But the view does not portray the trajectory's points inside the larger interval but outside the smaller as irrelevant to $v(t_0)$—since then it would deem every point other than $x(t_0)$ to be irrelevant, though the body's having $v(t_0)$ depends on the body's locations at instants other than t_0.

So on this reductive view, what does it take for a baseball's instantaneous velocity upon leaving a pitcher's hand to serve as an initial condition in a causal explanation of the ball's trajectory over the course of the succeeding Δt, during which the ball travels to home plate? No matter how small of a neighborhood surrounding t_0 we select, part of it is after t_0. Select a neighborhood and take a moment T in that part. For $v(t_0)$ to be a cause of the body's $x(T)$, where the body's possessing $v(t_0)$ is a relation's holding among the points in the body's trajectory in that neighborhood, our presupposition says that every relatum must be a cause of the body's $x(T)$. So $x(T)$ must be a cause of itself. This is problematic.

There may seem to be a way for this reductive view to negotiate the causal explanation problem. The body's trajectory during $<t_0, t_0 + \Delta t]$ is nothing but all of the ordered pairs (x, t) for moments in $<t_0, t_0 + \Delta t]$. For each moment T in $<t_0, t_0 + \Delta t]$, there is some neighborhood around t_0 that excludes T where the body's trajectory over that neighborhood suffices to fix $v(t_0)$. Perhaps, then, $v(t_0)$ can serve as an initial condition in causally explaining the body's trajectory over $<t_0, t_0 + \Delta t]$, even though $v(t_0)$ is nothing but a relation's holding among points in the body's trajectory, because each point in the body's trajectory in $<t_0, t_0 + \Delta t]$ can be causally explained by other points that suffice to fix $v(t_0)$. When the trajectory in $<t_0, t_0 + \Delta t]$ is explained point-by-point, no point in that trajectory has to help explain itself. In other words, an explanation of the body's trajectory in $<t_0, t_0 + \Delta t]$, with $v(t_0)$ as an initial condition, could be interpreted as follows: for any point $x(T)$ of that trajectory, there is a causal explanation having as one initial condition that a certain relation holds among the points in the body's trajectory in any neighborhood around t_0 that is small enough to exclude T.

Does this proposal respect our earlier presupposition: that a relation's holding is a cause only if the relata are? Not quite. On this proposal, the body's $v(t_0)$ is a cause of the body's $x(t_0 + \delta t)$ and is a relation's holding among the points in the body's trajectory in $[t_0 - \Delta t, t_0 + \Delta t]$. Yet not all of those relata are causes of the body's $x(t_0 + \delta t)$; only those $x(t)$'s where $t < (t_0 + \delta t)$ are causes. Of course, it might be suggested that when the relation involves a limit, the presupposition should not be required to apply fully, but only to this extent. Let's suppose that is right.

But $v(t_0)$ is supposed to be a cause of *all* of the points in the body's trajectory in $<t_0, t_0 + \Delta t]$. It is a *common* cause of the body's position at every later moment—no matter how remote from t_0, and certainly no matter how near. On the above proposal, the common cause we find in $v(t_0)$, a relation's holding among the points in the body's trajectory, is lost when we proceed to take the relata as causes. For any two points $x(t_0 + \Delta t)$ and $x(t_0 + \delta t)$ in the body's trajectory after t_0 (indeed, for any finite number of points), there is a neighborhood around t_0 where all of the points in the body's trajectory in that neighborhood, fixing $v(t_0)$, can serve as common causes of $x(t_0 + \Delta t)$ and $x(t_0 + \delta t)$. But no single neighborhood can play this common-causal role for *all* of the points in the body's trajectory in $<t_0, t_0 + \Delta t]$. So this reductive view fails to respect a slight extension of the principle we presupposed (that a relation's holding is a cause only if the relata are)—namely, that a relation's holding is a *common* cause only if the relata are.

This extension seems as plausible as the original principle. For example, on both occasions when we placed objects a and b on the balance, it tipped toward a. These two outcomes have a common cause: a's mass exceeding b's. This relation's holding is a common cause only because a's mass and b's mass, the relata, are common causes—just as our extended principle requires. Now suppose that there was a further occasion on which we placed object c along with b on the balance's right pan, and the balance still tipped to the left—toward object a. Is a's mass exceeding the sum of b's and c's masses now a common cause of the balance's having tipped toward a on all three occasions? I think not, and our extended principle agrees. For that relation's holding among a's, b's, and c's masses to be a common cause, according to our extended principle, those relata would all have to be causes of each outcome. But

c's mass was not a cause of the outcome on those occasions when c was not on the right pan.

On this reductive view, there is no stretch of the body's trajectory before and after t_0 that can play $v(t_0)$'s role as a cause of each point in the body's trajectory after t_0. It seems hopeless, then, to try to reduce $v(t_0)$ to a feature of the body's trajectory before and after t_0, since whatever $v(t_0)$ is reduced to must be a common cause of every point in the body's trajectory after t_0.

However, this problem may be avoided by a slightly different reductive approach. Consider the ramp function:

$$R(t) = \begin{cases} 0 \text{ for } t \leq 0 \\ t \text{ for } t \geq 0. \end{cases}$$

$R(t)$ is continuous, but at $t = 0$, it is nondifferentiable. $R(t)$'s graph is "bent" at $t = 0$. $R(t)$'s behavior at $t = 0$ is captured by its derivatives "from below" and "from above." For any function $f(x)$, we can define its limit as x approaches a "from below" (also known as "from the left") by amending our earlier definition of "$\lim f(x) = L$ as x approaches a": where we had stipulated "whenever $0 < |x - a| < \delta$," we merely add the requirement that $x < a$. That is, whereas we had been taking a neighborhood of $x = a$ that extends for a distance δ above and δ below a, now in taking the limit *from below* we are considering only the neighborhood extending for a distance δ below a. Analogously, we may define a function's limit as x approaches a "from above" (also known as "from the right"). It is easily shown that f has a limit *simpliciter* at $x = a$ if and only if its limits there from above and from below exist and are equal. We may use the limits from below and above to define a function's derivative from below and derivative from above, respectively. At $t = 0$, R's derivative from below is zero and from above is 1. Since these are unequal, $R(t)$ has no derivative *simpliciter* at $t = 0$.

Let's try to avoid the causal explanation problem by reducing a body's instantaneous velocity to its trajectory's time-derivative *from below*.[44] Then apparently, the body's $v(t_0)$ can serve as an initial condition in explaining the body's trajectory over the interval $<t_0, t_0 + \Delta t]$. For each moment T in $<t_0, t_0 + \Delta t]$, $x(T)$ has a causal explanation with an initial condition consisting of a relation's holding among the points

in the body's trajectory in any neighborhood extending below (but not above) t_0. Since each of those neighborhoods extends only below t_0, each can serve for all moments in $<t_0, t_0 + \Delta t]$. So we have apparently found a way to reduce $v(t_0)$ to a feature of the body's trajectory while respecting $v(t_0)$'s role as a common cause of all of the points in the body's trajectory in $<t_0, t_0 + \Delta t].$[45]

But what if $v(t_0)$ and the momentary acceleration $a(t_0)$ are serving as *effects* and not (merely) as causes? Interpreting $v(t_0)$ and $a(t_0)$ as relations holding among the points in the body's trajectory at and before t_0 may be advantageous when $v(t_0)$ and $a(t_0)$ are serving as causes, but it is inconvenient when they are serving as effects. Consider a charged body at location x_0 within an electric field (and feeling no other forces). The body's $a(t_0)$ is caused by its mass and the electric force it feels (in accordance with Newton's second law of motion, $F = ma$). That force $F(t_0)$, in turn, is caused (in accordance with $F = qE$) by the field E at x_0 along with the body's possessing charge q and occupying x_0 at t_0. So the body's occupying x_0 at t_0 is a cause of its $a(t_0)$. But according to the derivative-from-below proposal, $a(t_0)$ is a relation's holding among the points in the body's trajectory in an interval from some earlier moment up to and including t_0. I suggested earlier that a relation's holding is a cause only if each of the relata is. I now suggest that analogously, a relation's holding is an effect only if at least one relatum is. (If no relatum is affected, then how can their relationship have been affected? For example, suppose that a cause of my current weight's being below yours is my having recently followed a strict diet. Then my dieting must have affected either my current weight or yours.[46]) So for the body's occupying x_0 at t_0 to be a cause of $a(t_0)$, on the derivative-from-below proposal, the body's occupying x_0 at t_0 must be a cause of some point in the body's trajectory in an interval ending at t_0. But the body's occupying x_0 at t_0 cannot cause itself, and the other trajectory points in that interval occur *before* t_0 so it cannot cause them.[47]

This problem arose from our considering the body's instantaneous state of motion as an effect, whereas we had originally concentrated on it as a cause of the body's subsequent trajectory. Of course, we might try to avoid this problem by positing *two* instantaneous velocities and *two* instantaneous accelerations, one property of each pair involving the trajectory's time-derivative at t_0 from below and the other involving its derivative from above. The belows are the causes; the aboves are the effects. For example, the trajectory's first derivative with respect to time, taken from below at t_0, is the $v(t_0)$ that serves as an initial condition—a cause of the body's trajectory in a subsequent interval $<t_0, t_0 + \Delta t]$. The trajectory's second time-derivative from above at t_0 is the $a(t_0)$ that serves as an effect of the body's occupying x_0 at t_0, the electric field there then, and the body's mass and charge. On this view, talk of "$v(t_0)$" (or "$a(t_0)$") is ambiguous.[48]

One problem with the two-property view involves which of the two velocity properties should figure in other properties to which velocity is tied, such as kinetic energy ($\frac{1}{2}mv^2$) and momentum (mv). Perhaps these properties come in pairs as well, necessitating at least a doubling of all properties in which these, in turn, figure. This is fairly odd, though perhaps we could bear it.

A more serious problem with the two-property view is that if $v(t_0)$ as effect is distinct from (even if often equal to) $v(t_0)$ as cause, then there exists no single causal chain running *through* $v(t_0)$—that is, with the same $v(t_0)$ as effect *and* as cause. It cannot then be, for example, that the body's trajectory in $[t_0 - \Delta t, t_0>$ helped to cause its $v(t_0)$ which, in turn, helped to cause its trajectory in $<t_0, t_0 + \Delta t]$. Likewise, if there are two acceleration properties, then it is not the case that the force on a charged body causes its acceleration which, in turn, affects its electric field.[49] Rather, no common node joins the two halves of this causal chain. This seems like an unacceptably high price to pay.

Before I present my solution to the causal explanation problem (which analyzes facts about instantaneous rates of change as irreducibly subjunctive), let's look briefly at an alternative to the two reductive approaches that I have just examined.

According to "velocity primitivism," a body's instantaneous velocity is a property over and above its trajectory. The two are metaphysically independent, though connected by natural laws that give velocity its explanatory and causal roles. A body's velocity is one of its ontologically primitive properties—alongside mass, electric charge, and so forth.[50]

A common complaint against primitivism is that it fails to respect ontological parsimony.[51] That is, it adds to our ontology a property that we can dispense with (having already admitted trajectories) and a law reflecting this property's dispensability (since the law sets the new quantity equal to the trajectory's time-derivative—at least when things are well behaved). However, this objection is answered by the causal explanation problem. The additional property seems indispensable, since velocity as construed reductively cannot stand in the requisite causal and explanatory relations.

Nevertheless, I think that there is something fundamentally mistaken about primitivism: it fails to do justice to the fact that velocity is *essentially* something to do with trajectory (and nothing more than that). Unlike mass and electric charge, velocity is an essentially kinematic property. Trajectory, velocity, and acceleration belong to the same "family" of properties; velocity's place in this family is essential to it. The same considerations apply to other rates of change. Darkening cannot be something metaphysically distinct from anything involving degrees of darkness; cooling cannot be something metaphysically independent from temperature. Although the instantaneous rate of change of a body's trajectory is not *reducible* to some mathematical feature of that trajectory, certain relations among trajectory, velocity, and acceleration are metaphysical necessities rather than mere natural necessities (as velocity primitivism takes them to be).[52] For example, it is a broadly logical truth that if a body has a well-defined position, velocity, and acceleration throughout the interval $[t_0, t_0 + \Delta t]$, then

$$v(t_0 + \Delta t) = v(t_0) + \int_{t_0}^{t_0 + \Delta t} a(t)\ dt$$

and

$$x(t_0 + \Delta t) = x(t_0) + \int_{t_0}^{t_0 + \Delta t} v(t)\ dt.$$

These relations express the way that acceleration, velocity, and trajectory essentially belong to the same "family" of properties. Within this family, acceleration's relation to velocity is just like velocity's relation to position. These relations (along with Newton's second law of motion) are used to explain a body's trajectory during $<t_0, t_0 + \Delta t]$ by appealing to the net force on the body during $[t_0, t_0 + \Delta t]$, the body's mass, and initial conditions $x(t_0)$ and $v(t_0)$. This is the scientific explanation that originally prompted the causal explanation problem.

We might worry that if these relations are metaphysically necessary, but every scientific explanation requires contingent laws of nature, then the putative scientific explanations that prompted the causal explanation problem cannot be genuine scientific explanations. However, for $v(t_0)$ to help explain the body's subsequent trajectory, the body must *have* a subsequent trajectory; it must continue to exist. Given the initial conditions, a contingent law of nature is needed to explain why the body continues to exist. This part of the explanation is implicit in using a body's velocity during $[t_0, t_0 + \Delta t]$, its $x(t_0)$, and the above relation between position and velocity to explain a body's $x(t_0 + \Delta t)$ since, in appealing to the body's velocity during $[t_0, t_0 + \Delta t]$, we imply that the body exists throughout this interval. Analogous considerations apply to any scientific explanation in which a quantity's value at t_0 and its instantaneous rates of change through $[t_0, t_0 + \Delta t]$ explain its value at $t_0 + \Delta t$.

The idea that velocity is essentially kinematic is distinct from ontological parsimony and, I believe, just as strong of a motivation behind views that aim to reduce velocity to some mathematical function of trajectory. Let's now see how to respect velocity's essentially kinematic character without adopting a reductive view. The resulting account will solve the causal explanation problem by positing some primitive subjunctive facts.

4.6. Et in Arcadia Ego

That a body possesses a certain classical instantaneous velocity is (I propose) an irreducibly subjunctive fact. In particular:

> For a body at t_0 to have an instantaneous speed of V centimeters per second is for the body to exist at t_0 and for it to be the case that were the body to exist after t_0, the body's trajectory would have a time-derivative from above at t_0 equal to V cm/s.

In what context is this subjunctive conditional to be entertained? The sort of context where—regarding a fair coin tossed at t_0 and landing heads—the subjunctive conditional "If it were tossed at t_0, it would land heads" is false, since although the coin might land heads (and in fact did), it would have an equally good chance of landing tails. In other words, the subjunctive conditional is to be entertained in a context where a possible world's "closeness" to the actual world is not influenced by its sharing the actual post-t_0 outcomes of chance processes.[53]

I propose that acceleration be understood in nearly the same way:

> For a body at t_0 to have an instantaneous acceleration of A cm/s^2 is for the body to exist at t_0 and for it to be the case that the body's trajectory would have a second time-derivative from above at t_0 equal to A cm/s^2, were the body to exist for some finite temporal interval after t_0 *and* to experience then exactly the same net force as it actually experiences at t_0.

This last condition is required to accommodate the fact that in classical physics, a force field's nonzero region may have a sharp boundary—as when the electric field is zero inside and nonzero at and outside of a hollow, charged, spherical conducting shell. Consider a charged body on the shell but heading inside of it. The body has nonzero $a(t_0)$ but, were it to continue to exist, its trajectory's second time-derivative from above at t_0 would (neglecting nonelectrical influences) be zero, since it would thenceforth be in a region of zero electric field. However, were it to continue to exist while experiencing the same net force as it actually experiences at t_0, its trajectory's second time-derivative from above at t_0 would be nonzero.

No similar condition needs to be added to the definition of velocity because in classical physics, a discontinuity in a field produces a discontinuity in acceleration, not in velocity.[54] In a possible world operating according to classical physics (interpreted causally), outside influences act directly not upon a body's trajectory or upon its instantaneous velocity, but upon its velocity's rate of change. This suggests an important distinction between two reasons why quantities change. Some quantities change because something sets them at a new value (like a thermostat being reset). Other quantities change because something sets their rate of change (or the rate of change of their rate of change, or...) to a nonzero value. A rate of change (such as acceleration) that is directly set by outside influences must be understood in terms of a slightly different sort of subjunctive fact than a rate of change (such as velocity) that changes exclusively because outside factors directly set *its* rate of change. (I'll return to this shortly.)

My proposal does justice to velocity's essentially kinematic character since the relation between a body's $v(t_0)$, its still existing after t_0, and its trajectory's time-derivative from above at t_0 is metaphysically rather than merely naturally necessary. This does not keep the body's $v(t_0)$ from being a cause of the body's subsequent trajectory. When the trajectory's time-derivative from above at t_0 is explained by $v(t_0)$ and other conditions at t_0, these conditions do not include the body's still existing after t_0. Rather, they include the facts that combine with laws to explain why the body continues to exist after t_0. So when the trajectory's time-derivative from above at t_0 is explained by $v(t_0)$ and other initial conditions, the explanation goes through laws and not solely through metaphysical necessities.

On my proposal, a body that exists at t_0 and before, but not subsequently, can have a well-defined $v(t_0)$ despite its trajectory's having no time-derivative (from above or *simpliciter*) at t_0, since it may nevertheless have a certain kind of "potential trajectory" after t_0. Likewise, a body that exists at t_0 and subsequently, but not before, can have a well-defined $v(t_0)$ despite its trajectory's having no time-derivative from below (or *simpliciter*) at t_0.[55]

Consider a possible world where a body's sequence of positions is determined randomly; each point of the trajectory is the outcome of a separate random process. Trajectories in this world tend to be highly discontinuous, but consider a body that happens to move along a differentiable path over the interval $<t_0 - \Delta t, t_0 + \Delta t>$. On the two reductive accounts I explored earlier, the body has a well-defined $v(t_0)$, whereas according to velocity primitivism, it does not (since velocity's explanatory role is essential to it, but in this possible world, the body's trajectory is not explained in this way).[56] On my proposal, the body lacks a well-defined $v(t_0)$, since it is false (in the relevant sort of context) that were the body to exist after t_0, then its trajectory at t_0 would have a well-defined time-derivative from above. It might have, but it might not have, depending upon the outcome of a random process.[57]

In a possible world operating according to classical physics (interpreted causally), forces influence a body's trajectory by setting its instantaneous acceleration (via $F = ma$). Its instantaneous acceleration, in turn, affects its trajectory by affecting its instantaneous velocity. In contrast, in a possible world where the body's sequence of positions is determined randomly, the outcomes of cosmic die tosses (as it were) set the trajectory *directly*. Earlier I contrasted two reasons why quantities change. A body's $a(t_0)$ in a classical-physics world and $x(t_0)$ in a random-trajectory world are subject to influences that set them directly to new values irrespective of their former values. In contrast, when classical velocity changes, its new value depends on its former value; outside influences act directly only on its rate of change. Likewise, a body's temperature at t_1 results from its temperature at t_0 and its instantaneous rate of temperature change at each moment during $[t_0, t_1]$, where a cause of that rate at any moment is the instantaneous rate at which heat energy is entering the body, which in turn is caused by the difference between the body's temperature and the temperature of its surroundings. The incoming heat energy does not set the body's new temperature directly, but merely causes its temperature to rise.

If a quantity changes in the former fashion (by being set directly), then that quantity may have no instantaneous rate of change. But to change in the latter fashion, a quantity must have an instantaneous rate of change; the requisite subjunctive facts must exist. Indeed, for any quantity that changes in the latter fashion, the causal explanation problem arises. For example, if the body's instantaneous rate of temperature change at t_0 is a cause of the subsequent "trajectory" taken by the body's temperature, then that rate cannot be a relation's holding among the trajectory's points in an interval surrounding t_0. Moreover, the instantaneous rate of temperature change at t_0 is not only a cause, but also an effect. If the temperature difference at t_0 between the body and its surroundings is a cause of the body's instantaneous rate of temperature change at t_0, then that rate cannot be the temperature trajectory's time-derivative from below. The same issues that I have explored in connection with classical instantaneous velocity thus arise in connection with other instantaneous rates of change.

How, then, does my proposal resolve the causal explanation problem? As illustrated by the possible world where a body's sequence of positions is determined randomly, the body's actual trajectory after t_0 is not what makes it the case that the body would pursue a certain trajectory after t_0, were the body to exist after t_0. On my proposal, a body's $v(t_0)$ is not a relation's holding among points in the body's actual trajectory in an interval around t_0. So when the body's $v(t_0)$ serves as an initial condition in causally explaining each point in the body's subsequent trajectory, there is no danger of any of those points being a cause of itself and there is no obstacle to $v(t_0)$ serving as a common cause of each of those points.

On my view, facts ascribing classical instantaneous velocities (and certain other instantaneous rates of change) are subjunctive facts. In classical physics, a body's instantaneous velocity is an irreducible, fundamental component of its instantaneous state. The subjunctive fact cashing out a body's instantaneous velocity is ontological bedrock; it has no categorical ground.

Thus, irreducibly subjunctive facts lie at the heart of the mechanical philosophy—the worldview that inspires the urge to reduce all subjunctive facts to nonsubjunctive ones. Even in the Humean heartland beloved of those favoring such slogans as "All truths are made true by what is," excavation uncovers some ontologically primitive subjunctive facts. It would be parsimonious for the same sort of facts to serve as the lawmakers as well.

4.7. The Rule of Law

Since this book is approaching its close, we should finally take up the remark from Michael Faraday that served as one of its epigraphs: "The beauty of electricity, or of any other force, is not that the power is mysterious and unexpected, touching every sense at unawares in turn, but that it is under *law*...."[58] In 1858, when Faraday made this remark, the laws of electricity had not all been discovered. Nevertheless, Faraday was confident that all electric phenomena are covered by laws and that all other forces are, too—indeed, apparently, that laws cover every kind of situation that every possible kind of thing can get into. The laws are not just exceptionless, but also "complete"; there are no gaps in their coverage. Nothing falls beyond their sovereignty; nothing operates outside their jurisdiction.

But metaphors aside, what does the laws' "completeness" amount to?

A fact "covered" by the laws need not follow logically from the laws alone. To entail the fact being explained, the laws may need to be coupled with accidental "initial conditions." If the laws are statistical, then even the laws together with initial conditions may not suffice to entail some fact that they explain.

What would a *gap* in the laws' coverage be? For simplicity's sake, let's suppose that the laws specify (i) that everything consists entirely of elementary particles of certain kinds (A-ons, B-ons, and so forth); (ii) how elementary particles behave when they are not undergoing any interaction; (iii) the chances that various kinds of particles in various circumstances will interact and, if they do, the chances of various results—and nothing more. Suppose there is a gap in the laws' coverage: the laws entail nothing about the result of an A-on's interacting with a B-on when the two particles are separated by between 1 and 2 nanometers (nm). No law prohibits A-ons from being 1–2 nm from B-ons. Indeed, the laws specify the chance that an A-on interacts with a B-on if they are 1–2 nm apart—just nothing about what might result from such an interaction (not even the chances of various outcomes). To accentuate this gap in the laws' coverage, let's add that the laws do specify the chances of various outcomes when an A-B interaction occurs at less than 1 nm or greater than 2 nm. The only gap is between 1 and 2 nm.

The laws are then incomplete because they fail to cover an A-B interaction at 1–2 nm. The laws would not explain the outcome of any such interaction. Whether any such interaction ever actually occurs is irrelevant to the laws' incompleteness; it suffices that such an interaction is naturally possible.

Accordingly, let's say that the laws are "complete" exactly when in any history allowed by the laws, every event (except perhaps an event occurring at the universe's first moment) has a covering-law explanation. Provisionally (to be reconsidered near the end of the next section), the laws are "complete" if and only if

given any hypothetical world-history allowed by the laws (that is, where Λ is true) and

given any hypothetical event E (letting e be that E occurs) that does not concern the universe's first moment,

the history contains certain events

not involving chances and

all occurring at or before some moment T preceding the time with which E is concerned[59]

such that one of the following is true (where h is that those events occur):

(h & Λ) logically entails e,

(h & Λ) logically entails $\sim e$, or

there is some N such that (h & Λ) logically entails $ch_T(e) = N$ (in other words, that at T, E's chance of occurring is N).[60]

Roughly speaking, the laws are complete exactly when for any naturally possible history, every hypothetical event E's occurrence (or nonoccurrence) in that history is explained by the laws, where the explanation involves the laws together with certain prior conditions in that history entailing E's occurrence (or nonoccurrence) or at least E's chance at T.[61] The laws are complete exactly when they cover every actual event (except those occurring at the first moment) and this broad coverage is no accident.

According to a famous quip, on an English conception of (civil and criminal) law, everything is permitted that is not expressly forbidden, whereas on a Prussian conception, everything is forbidden that is not expressly permitted.[62] The two conceptions involve different defaults. But if the natural laws are complete, then everything (as far as sub-nomic matters are concerned) is either expressly forbidden or expressly permitted. There is no default; the laws plus some history through a given moment expressly categorize every hypothetical E. If the laws are complete, then they are both English and Prussian.

As Faraday's remark suggests, science appears to presume that the laws are complete. When scientists discover a new phenomenon, they try to find the laws covering it. That various proposed covering-law explanations have failed is not regarded as confirming that the phenomenon is governed by no laws at all. As Feynman notes, if ESP were verified, then since ESP is not a consequence of the known laws, its discovery would lead physicists to seek further laws governing ESP, not to take seriously the possibility that no laws cover it.[63]

If the laws are actually complete, then is their completeness just a notable feature of the actual universe or metaphysically compulsory? Of course, we could also ask whether it is metaphysically compulsory that some natural laws exist rather than none at all. Let's keep these two questions separate by asking: *If* there are laws (besides the broadly logical truths), must they be complete?

The natural laws are sometimes characterized as the rules of the "game" played by the universe's various inhabitants (particles, fields, or whatever).[64] Of course, the rules of a typical game are not complete in the same sense as the natural laws are presumed to be: they do not dictate the move that a player makes (or specify the chance of the player's making a certain move) in a given situation. But the game's rules are supposed to govern play in that they are supposed to specify, for any circumstances that could arise, which moves are allowed in that circumstance and which are not. Imagine a game with pieces that are moved around on a board. Perhaps one rule of the game says that a "fortress" can move diagonally when it is next to a "cardinal." Suppose that through a sequence of lawful moves beginning from a lawful initial arrangement of pieces, a fortress can find itself no longer next to a cardinal, but the rules specify nothing about the directions it can move in that situation. The game is then fundamentally flawed, its rules incomplete. But the natural laws, as nature's rules, are not supposed to be flawed in this way.[65]

A related metaphor understands the natural laws as "the software of the universe," directing the functioning of the hardware (particles, fields, or whatever).[66] Incomplete laws would then be analogous to a computer program afflicted with a "bug": it calls upon a subroutine that isn't there. In our earlier example where A-B interactions at $1-2\,nm$ fall into a gap, the cosmic software contains code (in BASIC!) something like the following:

⋮

540   REM I=0 MEANS NO INTERACTION, R IS THE
          SEPARATION
550   IF I=0 GOTO 1200
560   IF R < 1 GOTO 1300
570   IF R ≥ 1 AND R ≤ 2 GOTO 1400

⋮

But there is no step 1400. The program is incomplete. Unlike a computer, the universe cannot "crash," so the cosmic software cannot contain such a "bug." Insofar as the laws are "the software of the universe," the laws must be complete.

A widespread belief that the laws must be complete may also influence how science approaches singularities: situations where a physical quantity figuring in putative laws goes undefined, preventing the "laws" from yielding physically meaningful predictions (even statistical ones) regarding those situations. For example, according to classical electromagnetic theory, it is a law that the electric field at a given location x_0 is equal to the vector sum of contributions from all of the universe's charges, where (roughly) each charge's contribution is proportional to the reciprocal of the square of the charge's distance from x_0 and directed from x_0 away from the charge. Hence, if there is a charged point body at x_0, then its contribution is infinite and has no well-defined direction. Hence, the "law" breaks down for point charges. It has a gap there.

However, physicists generally regard singularities as indicating not that the "laws" are incomplete, but rather that they are not genuinely laws. Perhaps quantum mechanics or its successor must be used to determine a point charge's contribution to the electric field at its location—or to prohibit point charges.

This attitude toward singularities seems to have been Einstein's. According to one of his collaborators, "It seems Einstein always was of the opinion that singularities in a classical field theory are intolerable...because a singular region represents a breakdown of the postulated laws of nature. I think that one can turn this argument around and say that a theory that involves singularities and involves them unavoidably, moreover, carries within itself the seeds of its own destruction...."[67] This attitude is quite common[68] and motivates Penrose's and Hawking's "cosmic censorship" hypothesis, a version of which is roughly that there are no naked singularities (that is, no singularities that are able to affect the outside universe because they are neither at the end of time nor safely hidden behind event horizons) other than perhaps at the beginning of time. The laws of physics say nothing about what a naked singularity would spew forth into the rest of the universe, nothing even about its chances of emitting various things. As John Earman remarks: "The principles of classical GTR [general theory of relativity] do not tell us whether a naked singularity will passively absorb whatever falls into it or will regurgitate helter-skelter TV sets, green slime, or God only knows what."[69] Clearly, then, the natural possibility of a naked singularity would make the laws incomplete.[70] Earman again: "Perhaps one can also argue that violations of cosmic censorship would show that classical GTR is incomplete in a stronger sense. The premise required is not that determinism holds but the weaker premise that all physical processes be law governed. The argument would be completed by showing that classical GTR places no constraints, not even statistical ones, on what can emerge from a naked singularity."[71] With the laws' "completeness," I have tried to capture the premise "that all physical processes must be law governed" if there are laws at all.[72]

However, familiar accounts of natural law fail to entail that the laws (if there are any) must be complete. For example, Lewis's "Best System" need not be complete. Although a complete system would be mighty informative, it need not be simple. An incomplete system would be less informative but could be so much simpler as to make it better than any complete system. There would then be no profit in filling its gaps. For example, suppose there exist many bodies of "gas," each characterized by two fundamental quantities, P and V. Suppose that over the universe's history, many have $P \leq 500$ or $V \leq 500$ (in some units). For each, $P = V$. There are only a few where $P > 500$ and $V > 500$. They do not obey $P = V$; they fall into no simple pattern. A complicated curve could be fit through them, but a deductive system would (in the absence of other considerations) be better off including "$P = V$ for $P \leq 500$ or $V \leq 500$" and leaving unfilled the gap above this threshold.[73]

Armstrong's account of laws as relations of "nomic necessitation" among universals likewise fails to entail that the laws must be complete. Even given the universals that exist, it is not metaphysically compulsory that their nomic-necessitation relations be rich enough to make the laws complete.[74]

My proposal does better at accounting for the laws' completeness—once subjunctive facts join the sub-nomic facts at the bottom of the world.

4.8. Why the Laws Must Be Complete

Let's start with a quick-and-dirty explanation. Suppose, for the sake of *reductio,* that there is a gap in the laws' coverage. Return to my example where the laws fail to cover the outcomes of A-B interactions at 1–2 nm, though the laws not only allow such interactions but also specify the chances of various results of A-B interactions at less than 1 nm or greater than 2 nm. Suppose that in a given spatiotemporal region L, no A-B interactions actually occur at 1–2 nm. Consider ϕ:

> Had an A-B interaction occurred in L at a distance of 1–2 nm, then all such interactions would have turned the interacting particles into green slime.

Since the laws Λ are silent about the possible result of an A-B interaction at 1–2 nm, ϕ is logically consistent with $\Lambda°$ (the Stable set containing Λ's members along with various subjunctive conditionals, including those ensuring Λ's sub-nomic stability—as I discussed in section 4.3).

Hence, for $\Lambda°$ to be Stable, $\Lambda°$'s members must be preserved under the supposition that ϕ holds.

However, they would not be preserved. Suppose that green slime is *wildly* different from what the laws say results from A-B interactions at less than 1 nm or greater than 2 nm. Had it been the case that ϕ, then presumably the laws governing A-B interactions at other distances would (or at least might) have been different—in particular, have assigned chances to something like green slime being produced in interactions at those distances. Had it been the case that green slime would have been produced by all A-B interactions in L at 1–2 nm (had some such interactions occurred), then green slime would perhaps also have been produced by some A-B interactions at other distances (or in other space-time regions). That is:

[(An A-B interaction occurs in L at 1–2 nm) $\Box\!\rightarrow$
(green slime is produced by all such interactions)]
$\Diamond\!\rightarrow$
(some A-B interactions at other distances also produce green slime).

But the laws are violated if some A-B interactions at other distances produce green slime.

Of course, A-B interactions at 1–2 nm *could* be very different from A-B interactions at other distances; there is no metaphysical obligation for the laws to vary "smoothly" with distance. It suffices for my argument that in certain contexts, the laws governing A-B interactions at other distances *might* have been different, had it been the case that green slime would have been produced by all A-B interactions in L at 1–2 nm (had there been any such interactions).

This seems very plausible. After all, consider this counterfactual conditional:

[(An A-B interaction occurs at greater than 2 nm) $\Box\!\rightarrow$
(green slime is produced by all such interactions)]
$\Diamond\!\rightarrow$
(some A-B interactions at other distances also produce green slime).

This counterfactual conditional seems true. Admittedly, its truth does not threaten $\Lambda°$'s Stability; its antecedent posits the truth of a counterfactual conditional that conflicts with the conditionals in $\Lambda°$, and so $\Lambda°$ does not need to be preserved under this antecedent in order to be Stable. But that $\Lambda°$ is not preserved under this antecedent (involving A-B interactions beyond 2 nm yielding green slime) suggests that $\Lambda°$ also fails to be preserved under the supposition that green slime would have been produced by all A-B interactions in L at 1–2 nm (had there been any).

To summarize: Suppose (for the sake of *reductio*) that the laws Λ are incomplete. Consider a counterfactual conditional ϕ specifying that something wild would have happened, had there been a case falling into the gap. Had ϕ, would the laws still have held? I have argued that the answer is not "Yes." But the answer must be "Yes" for $\Lambda°$ to be Stable.[75] Hence, $\Lambda°$ is not Stable, and so (by my earlier argument) Λ is not the set of laws. Contradiction. So the laws must be complete.

That was the quick-and-dirty argument. Now for a slightly more careful version—in two steps.

First step: Plausibly, if $p \,\square\!\!\rightarrow q$ holds, where p is false but logically consistent with the actual laws Λ, then one of two options must hold:

(i) $(p \,\&\, \Lambda)$ logically entails q, or

(ii) $(p \,\&\, \Lambda)$ does not logically entail q, but there is some actual sub-nomic fact f that is not a law and that the context implicitly invokes where $(p \,\&\, \Lambda \,\&\, f)$ logically entails q.

For example, if $p \,\square\!\!\rightarrow q$ is that had I struck the match, it would have lit, then in a typical case where this counterfactual conditional is true, f is that the match is dry, well-made, surrounded by oxygen, and so forth. These two options (and the match example) are suggested by Goodman's famous examination of counterfactuals, to which I alluded earlier. On the second "Goodman option," it may even be that q is logically entailed by $(p \,\&\, f)$—without the aid of Λ. For example, suppose that whether a given radioactive atom decays is governed only by irreducibly statistical laws. Suppose that the atom actually does decay, but I made a bet that it wouldn't and so lost. Had I bet that it would decay, then I would have won. (That is true in some contexts, but not all.) Here typically a suitable f is that the atom decays, and q is logically entailed by $(p \,\&\, f)$. Likewise, in Goodman's match example, context requires us to "hold fixed" various features of the match's state, so if $p \,\square\!\!\rightarrow q$ is that the match would have been surrounded by oxygen, had it been struck, then q is logically entailed by f—without the aid of Λ.

Not every counterfactual conditional that is contingently true must be "covered" by a law.[76]

Goodman's point applies more broadly: not only must one of the two Goodman options hold if $p \;\square\!\!\rightarrow q$ is in fact true, but also one of them *would* have held *had* $p \;\square\!\!\rightarrow q$ been true. For example, suppose that in fact, the match is wet, so "struck $\square\!\!\rightarrow$ lit" is false. Under typical conditions, had "struck $\square\!\!\rightarrow$ lit" been true (and the match not been struck), then there would have been laws Γ and a salient accidental truth f (that the match is dry, oxygenated, and so forth) such that "The match lights" is entailed by "The match is struck" together with f and Γ. Accordingly, in general: Had p been false and $p \;\square\!\!\rightarrow q$ held, where p is logically consistent with whatever the laws Γ would have been had p been false and $p \;\square\!\!\rightarrow q$ held, then one of these two "Goodman options" would have held:

(*i*) (p & Γ) logically entails q,

(*ii*) (p & Γ) does not logically entail q, but there is some subnomic claim f that would have held accidentally, had $\sim p$ and $p \;\square\!\!\rightarrow q$ held, and that the context implicitly invokes, where (p & Γ & f) logically entails q.

Now which option applies if the laws have a gap for A–B interactions at 1–2 nm and $p \;\square\!\!\rightarrow q$ is our earlier ϕ ("Had an A–B interaction occurred in L at a distance of 1–2 nm, then all such interactions would have turned the interacting particles into green slime")? Had ϕ held (and—let this qualification henceforth be tacit—its antecedent been false), which Goodman option would have applied?

If the first Goodman option would have applied, then had ϕ held, the laws would have included some or another "green-slime law" (such as that every A–B interaction at 1–2 nm produces green slime). That is: $\phi \;\square\!\!\rightarrow$ a green-slime law.

If the second Goodman option would have applied, then had ϕ held, ϕ would have held partly by the grace of some f. But suppose we amend ϕ to something like "Had an A–B interaction occurred in L at 1–2 nm *prior to which there was nothing in the universe's entire history except for an A-on and a B-on approaching each other*, then all such interactions would have turned the interacting particles into green slime." We might even imagine adding further to the italicized portion of the antecedent, making it completely describe the posited universe's subnomic history until the A–B interaction takes place. This counterfactual's antecedent (unlike "Had the match been struck") leaves no room for some f to supplement it in entailing its consequent. Therefore, had this ϕ held, then the first Goodman option would have applied to it: ($\phi \; \square\!\!\rightarrow$ green slime law).[77]

Second step: Is it the case that had ϕ held and the laws included a green-slime law, then Λ would still have held? I suggest not—at least, it is not the case that in every context, Λ would still have held. It seems very implausible that in every context, the laws governing A-B interactions at other distances would still have held, had there been a green-slime law for A-B interactions at 1–2 nm. There are plenty of examples where the actual laws would (or, at least, might well) not still have been true, had there been an additional law with which they are nevertheless logically consistent. For instance, had there been a law prohibiting gold cubes larger than a cubic mile, wouldn't the force laws perhaps have been different? The actual force laws *could* presumably still have held. (The laws could merely have imposed a further constraint on, say, the universe's possible initial conditions.) But I see no reason to insist that the actual force laws *would* still have held, had the gold-cubes fact been a law.

Let's now gather the fruits of this argument's two steps. Suppose (for the sake of *reductio*) that the fundamental physical laws Λ are incomplete (using the A-B example). Then

$\phi \; \square\!\!\rightarrow$ green-slime law
(ϕ & green-slime law) $\diamond\!\!\rightarrow \sim\!\Lambda$ [at least in some context]

Therefore, by the kind of transitivity that counterfactuals respect,

$\phi \; \diamond\!\!\rightarrow \sim\!\Lambda$ [at least in some context]

contrary to $\Lambda°$'s Stability. *Reductio* achieved.[78]

Let's look at this argument in one final way. Suppose there is a law m specifying that any "coin flip" has a 50% chance of yielding heads and a 50% chance of yielding tails. In certain contexts, it is true that had it been the case that the flip would have landed heads had I flipped the coin once in L, then m would not still have held.[79] Indeed, even coin flips outside of L would (or, at least, might) not still have had 50% chances of yielding tails. Just as the supposition of the subjunctive fact "flip in L $\square\!\!\rightarrow$ heads" posits that such a flip's outcome is (as it were) preordained, so likewise the supposition that "Were there an A-B inter-action at 1–2 nm, all such interactions would produce green slime" (ϕ) posits that such an interaction's outcome is preordained. Just as if it had been the case that "flip in L $\square\rightarrow$ heads," then coin flips outside L would (or, at least, might) not still have had 50% chances of yielding tails, so likewise had ϕ, then the actual laws about A-B interactions at other distances would (or, at least, might) not still have held.

I have concluded that the laws (if there are any besides the broadly logical truths) must be complete. "What colossal presumption," you may say, "for a philosopher reasoning a priori to purport to ascertain such a contingent fact about the universe!" You might press the point:

> Surely it should be left for empirical science to figure out what the laws happen to be like. In particular, we should not pre-judge the ways that the laws might constrain the future given the past. The requirement that the laws be "complete" auda-ciously presumes that the laws could constrain some hypothet-ical future event E given past events only by entailing e, entailing $\sim e$, or entailing E's chance. But there is no a priori limit to discovery in science. We might someday discover addi-tional ways for laws to explain sub-nomic facts—ways that make the laws "incomplete" in the precise sense formulated above, but that do not intuitively involve a gap in the laws' coverage. After all, had this book been written before quantum mechanics was discovered, it might have posited a "complete-ness requirement" that left no room for irreducibly statistical explanations and ontologically primitive chances in funda-mental physics. How can we be sure that any purported "com-pleteness requirement" has allowed for *all* of the ways that laws *could* (as a matter of natural possibility, let alone metaphysical possibility) "cover" an event?

I agree with this objection. I suggest that no particular "completeness requirement" is metaphysically necessary; the requirement differs in different possible worlds. In some possible worlds where none of the fundamental laws ascribes chances, any violation of determinism would constitute a gap in the laws' coverage. In other possible worlds where there are also statistical laws, the completeness requirement is the one given above.

Furthermore, consider a universe where some chance processes have only *chances* as their outcomes. Such a process has various chances of yielding various outcomes consisting exclusively of various chances of e. Until a given process has run its course (at T'), there is no N such that $\mathrm{ch}(e) = \mathrm{N}$. There is (at an earlier moment T) only some chance M that when the process later yields an outcome (at T'), e's chance will then be N. At T, there is only $\mathrm{ch}_T(\mathrm{ch}_{T'}(e) = \mathrm{N}) = \mathrm{M}$, an irreducibly second-order chance.[80] In such a universe, the completeness requirement is more liberal than the one given above. It is satisfied if the laws and various events (not involving chances, and occurring at or before some moment T preceding the time with which E is concerned) entail that some M is the chance at T that at some later moment T', E's chance is N. If the laws say nothing about the outcomes of A–B interactions at 1–2 nm, even about the chance of green slime's later having a given chance of ultimately resulting, then the laws are incomplete. The laws cannot contain such a gap, on pain of $\Lambda°$'s lacking Stability.

Although (I have argued) it is metaphysically compulsory that some or another "completeness requirement" hold, no particular completeness requirement is metaphysically compulsory. Rather, a particular completeness requirement holds in a given possible world as a meta-law—of the kind I discussed in chapter 3. (There I suggested in passing that in a deterministic universe, determinism might hold as a meta-law; it might be no mere "coincidence" that the laws are rich enough to determine the future given the past.) The meta-law expressing the completeness requirement specifies the manner in which every event must be "covered" by first-order laws. I suggest that in any possible world where there are laws, there must be an appropriate completeness principle holding with the modal force of a meta-law, constraining the laws in the same manner as symmetry meta-laws would.

Consider, for example, a possible world where the first-order laws include Newton's second law of motion and various force laws. As a meta-law, time-displacement symmetry would restrict the kinds of fundamental forces there could be. For example, it would preclude a fundamental force law demanding that all bodies feel a component force in a given direction that is zero until a certain time T and a constant nonzero strength thereafter. Similarly, a completeness meta-law would restrict the kinds of fundamental forces there could be. For instance, it would preclude a fundamental force on a body varying in a given direction as the square-root of the body's speed in that direction. Such a force (if permitted to act in isolation on a body at rest) generates from Newton's second law an equation of motion that is satisfied by more than one trajectory: by the body's sitting still for *any* span of time, and then spontaneously beginning to move.[81] The body's launching into motion (or remaining at rest) is not covered by the laws; they do not even ascribe chances (given the prior history) to these events. Like a symmetry meta-law, a completeness meta-law explains why the first-order laws have a certain feature.

By allowing different possible law-governed worlds to have different completeness meta-laws, my view neither forecloses the conceptual innovations and empirical discoveries open to future science nor imposes a priori limits on the laws' ingenuity in constraining the future given the past. I have argued only that there must be *some* completeness meta-law suited to the actual laws. Empirical science has the task of discovering what it actually is.

4.9. Envoi: Am I Cheating?

Since the laws fail to supervene on the sub-nomic facts but do supervene on the subjunctive facts, it seems natural to try to enlist the subjunctive facts as the lawmakers. I have argued that this approach pays many dividends. Subjunctive facts are pretheoretically quite familiar— arguably more so even than facts about the natural laws. Nevertheless, to cast them as the lawmakers may seem vaguely like cheating.

When is it cheating to take a certain kind of fact as a primitive in a philosophical analysis of some other kind of fact? Goodman addresses this question in a section of *Fact, Fiction, and Forecast* aptly entitled "On the Philosophic Conscience." I love this passage so much that I cannot resist quoting it, as my parting gift to any reader who has made it this far:

> In life our problems often result from our indulgences; in philosophy they derive rather from our abnegations. Yet if life is not worthwhile without its enjoyments, philosophy hardly exists

without its restraints. A philosophic problem is a call to provide an adequate explanation in terms of an acceptable basis. If we are ready to tolerate everything as understood, there is nothing left to explain; while if we sourly refuse to take anything, even tentatively, as clear, no explanation can be given. What intrigues us as a problem, and what will satisfy us as a solution, will depend upon the line we draw between what is already clear and what needs to be clarified.[82]

Where, then, ought we to draw this line? We cannot know where it should be drawn until we have explored the results of drawing it in various places. We have taken too much as primitive if analyses of the rest are philosophically sterile. But we have taken too little as primitive if analysis of the rest is impossible. If our analyses of the rest turn out to be neither too easy nor too difficult, but instead to supply a host of elegant connections, explanations, unifications, and other unexpected payoffs, then we may have done well in our selection of primitives.

My proposal has to be judged by this standard. If the verdict is favorable enough, then it may overrule any initial inclination to avoid placing subjunctive facts among the primitives. Though my "philosophic conscience" pricks me somewhat whenever I propose treating subjunctive facts as the lawmakers, I have come to take this option seriously. In that spirit, I offer it to you.

Notes

1. This example originally appeared in Reichenbach 1947, on p. 368. By Reichenbach 1954, it had been promoted to pp. 10–11.
2. I shall use "fact" and "truth" interchangeably.
3. Hooker 1632: 8 (1.iii.4).
4. Hempel 1966: 56.
5. Braine 1972: 144.
6. OK, even Bill Gates could not afford a cubic *mile* of gold. He could afford a cubic meter of gold—or a cubic mile of good-quality timothy hay. But let's set these details aside for the sake of a vivid example.
7. To reduce clutter, I will typically omit quotation marks (and fussy corner quotes) around symbolic expressions where no confusion should result.
8. Quine (1960: 222) uses this example to illustrate the context-sensitivity of counterfactuals—as I will do in a moment.
9. See Goodman 1947, 1983.
10. Notably Lewis 1973, 1986b.
11. This idea has been familiar since Chisholm (1946) and Goodman (1947).
12. Of course, there may be conversational contexts where "Had copper been electrically insulating, then the wires on the table would not have been made of copper" is true. To demonstrate my point, it suffices that there is some context in which "Had copper been electrically insulating, then the wires on the table would still have been made of copper (but would not have been electrically conductive)" is true.

Perhaps the wires are *essentially* electrical wires, so that in every context, it is not the case that had copper been electrically insulating, then those wires would still have been made of copper. You may fiddle further with the example or concoct a different one. (For instance, had the gravitational force been stronger by 1 part in 10^{9000}, then the wires would still have been made of copper.)

Some philosophers maintain that "All copper is electrically conductive" is a broadly logical truth, instead of a law, on the grounds that part of being copper is being electrically conductive. These philosophers can construct an example similar to mine involving whatever facts they consider laws rather than broadly logical truths. In later chapters, I shall take issue with "scientific essentialism," according to which all laws are broadly logical truths.

13. This example is from Haavelmo (1944: 29), who says that g has "invariance with respect to certain hypothetical changes."

14. Had electric fields acted on electrons in a certain manner (running contrary to the actual laws), then copper would not have been electrically conductive—though for electric fields to operate in that manner is logically consistent with copper being conductive: If electric fields acted upon electrons in that unlawful manner, but there were other departures from the actual laws as well, then those other departures could compensate for the electric field's departure, thereby making copper electrically conductive once again. So the counterfactual supposition concerning electric fields is logically consistent with copper being electrically conductive (though logically inconsistent with other laws) and yet under that supposition, copper would not have been electrically conductive—just as the supposition that I strike the match is logically consistent with the match remaining unlit (since it is logically consistent with the match being wet), yet had I struck the match, it would have lit (since the match is actually dry).

15. When I say that $p \:\square\!\!\rightarrow m$ is true in a given context, I am not talking about whether the symbol "$p \:\square\!\!\rightarrow m$" in a given context stands for a truth. After all, what a given symbol stands for—indeed, whether it stands for anything at all—depends on accidental facts about the language being used. The laws would have been no different even if a given symbol had never stood for anything. As a matter of accidental fact, we use certain symbols to express certain propositions. Apparently we use a given symbol of the form "$p \:\square\!\!\rightarrow m$" (such as "Had I struck the match, it would have lit") to express different propositions in different contexts. According to NP, that certain of these propositions are true is related to the fact that m is a law. Roughly speaking, a given sentence of the form "$p \:\square\!\!\rightarrow m$" (an abstract, necessarily existing thing—not any particular token utterance, which might never have been made) has a certain (Kaplanian) character, and to any context (also an abstract, necessarily existing thing) in which that sentence is meaningful, that character assigns a certain proposition: what the sentence expresses in that context. When I say that m's lawhood is connected to the fact that in a given context, $p \:\square\!\!\rightarrow m$ is true, I mean to be connecting m's lawhood not to the accident that the symbol "$p \:\square\!\!\rightarrow m$" is used in a certain way, but rather to the truth of the proposition expressed in that context by the sentence $p \:\square\!\!\rightarrow m$.

16. For simplicity, I shall later sometimes omit some of these qualifications in stating NP and defining "sub-nomic stability."

Of course, in certain contexts we properly assert counterfactuals that we do not believe to be true—as when we are performing a play. NP does not concern which counterfactuals are properly *asserted* in certain contexts. It concerns which are *true* there.

Some contexts may be out of place in certain scientific fields (as in my example of emergency room medicine: in that field, there is no place for a context where "Had our vital organs been arranged differently" is relevant). Accordingly, we might investigate whether some field-relative version of NP holds:

m is a law *of a given scientific field* if and only if for any conversational context *that is relevant in that field* and for any *p* that is a relevant counterfactual antecedent in that context and logically consistent with all of the laws *of that field* (taken together), the proposition expressed in that context by "*p* □→ *m*" is true.

Elsewhere (Lange 2000, 2002b, 2004) I have pursued this line of thought in trying to understand what a law of (say) island biogeography or hydrodynamics would be and why such laws would be irreducible to the fundamental laws of physics. For the sake of simplicity, I shall not elaborate this line of thought in this book. I shall nearly always speak here only of the natural laws *simpliciter* (and of the laws having to be preserved in *all* conversational contexts—by which I do not mean merely in all contexts that happen actually to be realized). But my remarks can always be extended *mutatis mutandis* to cover a field-relative notion of lawhood.

Although a counterfactual conditional's truth-value is context-sensitive, lawhood (for a given scientific field) is not, since *m*'s lawhood (for a given scientific field) is associated with *m*'s preservation *in all contexts* (relevant in that field).

17. This example is from Leeds 2001: 193.

18. See my 2000: 201–6.

19. In chapter 3, I will explore an alternative strategy.

20. I take it that these do *not* include facts expressed by subjunctive conditionals. But I will return to this point in chapter 4, where I will try to avoid discriminating against these facts.

21. Or at least a law-*statement*. I trust that context will make it clear whether by a "law," I mean some fact or some claim expressing it or some proposition that it makes true. (In some previous publications, I used the term "non-nomic" instead of "sub-nomic." I now think that "sub-nomic" is less potentially misleading; my change of terminology has no other significance.)

22. The expression "The rock in your hand" could be understood as referring in any possible world to whatever unique thing is in your hand in that world (or as failing to refer, if there is no such thing in that world). Alternatively, this expression could be understood as referring in any possible world to

the rock that is *actually* in your hand (if that thing—or its counterpart—is in that world). That is, the expression could be understood as rigidly designating that individual thing. Suppose that the rock actually in your hand is an emerald, and that its emeraldhood is one of its essential properties; *that rock* could not exist without possessing that property. Then when "The rock in your hand" is understood rigidly, the laws (such as "All emeralds are green") are not preserved under the counterfactual supposition "Had the rock in your hand been yellow." So in order for NP to hold, it must not cover such counterfactual suppositions, so interpreted. Indeed, they do not qualify as "sub-nomic." The counterfactual antecedent "Had the rock in your hand been yellow," understood rigidly, is effectively "Had the rock *actually* in your hand been yellow," and sub-nomic claims contain no such modal elements. Proper names (such as "water" and "Lange"), though rigid, are not like "The rock in your hand" if the metaphysical (and hence, by courtesy, the natural) necessities include such truths as "Water is H_2O" and "Lange is human." But "Lange's favorite gem," though behaving rigidly, is excluded from sub-nomic claims.

As another example, suppose that Smith says, "All emeralds are yellow." Had Smith's remark been true, the laws of nature would have been different. But the expression "Smith's remark" was just now rigidly designating what Smith *actually* said, so this counterfactual conditional's truth is no threat to NP once NP is restricted to counterfactual suppositions that are "sub-nomic." (If we do not take the expression "Smith's remark" rigidly, then had Smith's remark been true, Smith would have to have said something different, in view of the laws of nature.)

In the next chapter, I give an account of what it is for a sub-nomic claim's truth to be necessary. If a sub-nomic claim included such rigid designators, then the necessity of such a claim's truth would be *de re* necessity; its necessity would be attributing modal properties to objects *simpliciter* rather than under certain descriptions. My concern throughout is exclusively *de dicto* necessity.

23. In sections 3.6 and 3.7, I will look more closely at the relation between laws of nature and facts about objective chances.

Other accounts of natural law may need to find some other way to carve out the facts that I am calling "sub-nomic." David Lewis's "Best System Account," for instance (which I will describe in chapter 2), deems facts about objective chances to be (in part) facts about the laws, so under Lewis's account, facts about chances do not qualify as "sub-nomic" by my definition. Presumably, the facts that I am calling "sub-nomic" could be picked out somehow by these other accounts. But the means of doing so would have to be tailored to the account.

Whether the "sub-nomic facts" include facts about the possession of various dispositions by various objects (such as that the vase is fragile, the rubber band is elastic, or the wire is electrically conductive) depends upon the nature of dispositional facts. That is controversial. If "The rubber band is elastic" is shorthand for a fact about the laws (e.g., "The rubber band possesses an intrinsic property such that it is a law that anything possessing that property while stretched exerts a restoring force"), then the dispositional fact is not sub-nomic. On the other hand, if "The rubber hand is elastic" is shorthand for a conjunction of subjunctive conditionals with sub-nomic antecedents and consequents (e.g., "Were the rubber band stretched without first being heated or placed in liquid nitrogen or anything like that, then it would exert a restoring force"— where it is tacitly understood which sub-nomic conditions qualify as "anything like that"), then the dispositional fact qualifies as sub-nomic (since it is governed by laws in the same manner as various nondispositional facts).

In that case, it might seem inconsistent for me to classify dispositional facts as sub-nomic while excluding the facts expressed by subjunctive conditionals with sub-nomic antecedents and consequents (see note 20). Ultimately, in chapter 4, I will try to treat the facts expressed by these subjunctive conditionals on a par with sub-nomic facts. By the same token, in some principles similar to NP, I will include not only counterfactuals like $p \ \square\!\!\rightarrow m$, but also "nested counterfactuals" such as $p \ \square\!\!\rightarrow (q \ \square\!\!\rightarrow m)$. If a dispositional fact is shorthand for a conjunction of counterfactuals with sub-nomic antecedents and consequents, then having included nested counterfactuals in principles similar to NP, we would seem to have no grounds for excluding a counterfactual with a consequent like "...then the rubber band would have been elastic."

Although in this chapter I will include some kinds of nested counterfactuals within the scope of some principles similar to NP, I will ultimately have to wait until chapter 4 to include all kinds of nested counterfactuals, as I think I should. I will argue there that the facts expressed by subjunctive conditionals with sub-nomic antecedents and consequents should be treated on a par with sub-nomic facts.

24. Although I accept that $\sim (p \ \lozenge\!\!\rightarrow \sim m)$ logically entails $(p \ \square\!\!\rightarrow m)$, I do not accept the reverse entailment. That is, unlike some philosophers, I do not regard "Had p obtained, then q might have obtained" as logically equivalent to "It is not the case that had p obtained, then $\sim q$ would have obtained." Although I believe that this "might = not-would-not" relation holds for some counterfactuals in some contexts, I think that in other cases, $p \ \lozenge\!\!\rightarrow q$ means "It is not the case that had p obtained, then $\sim q$ would *have to* have obtained." (In my 2000, I argue that a would-*have-to*-have conditional entails the corresponding would-have conditional, but the would-have does not entail the would-*have-to*-have. For example, had I gone out to lunch, I would have eaten Chinese food, but I wouldn't *have to* have; there are plenty of other restaurants around.) In yet other cases, $p \ \lozenge\!\!\rightarrow q$ may mean "Had p obtained, then q could have

obtained" or "Had p obtained, then it would have been possible for q to have obtained." (All of the "might"s I discuss are intended to involve alethic modalities, not an epistemic modality; see chapter 2.)

But these fine points will not play an important role in my argument. I occasionally use might-conditionals as intuition pumps—as when I said that there might have been a gold cube exceeding a cubic mile, had Bill Gates wanted one built, thereby suggesting it is not the case that there would still not have been a gold cube exceeding a cubic mile, had Bill Gates wanted one built. For these purposes, it suffices that at least in these cases, the "might = not-would-not" relation holds. Furthermore, under any of the above interpretations of might-conditionals, $\sim (p \Diamond\!\!\rightarrow q)$ entails $p \Box\!\!\rightarrow \sim q$ (which is the direction of entailment that I accept). For example, where "might = not-would-*have-to*-have-not" holds, $\sim (p \Diamond\!\!\rightarrow q)$ entails "Had p obtained, then $\sim q$ would *have to* have obtained," which entails "Had p obtained, then $\sim q$ would have obtained" $(p \Box\!\!\rightarrow \sim q)$.

I cannot accept that $(p \Box\!\!\rightarrow m)$ logically entails $\sim (p \Diamond\!\!\rightarrow \sim m)$, because otherwise I would have to accept that the logical truth $(p \,\&\, \sim p) \Box\!\!\rightarrow p$ logically entails $\sim ((p \,\&\, \sim p) \Diamond\!\!\rightarrow \sim p)$, which (by the second might-would relation I endorse in the main text) logically entails $\sim ((p \,\&\, \sim p) \Box\!\!\rightarrow \sim p)$, which is the negation of a logical truth. (Compare Lewis 1986b: 65.)

25. Bennett (2003: 168) gives the example "Had the leaves been dry, then it would have been dangerous to throw a match on them" in that had a match been thrown on them, a fire would have started. Bennett's characterization of his example as a "subjunctive conditional [that] has another as consequent" suggests that if nested counterfactuals are included in NP, then we have no grounds for keeping out counterfactuals involving dispositions—and vice versa. (Recall note 23.)

26. Faraday's entry 10,040 in his laboratory diary (March 19, 1849): "Nothing is too wonderful to be true, if it be consistent with the laws of nature." Likewise, Wigner: "all the...elements of the behavior [i.e., global world history] which are not specified by the laws of nature...can be chosen arbitrarily" (1997: 186–87).

27. The quoted passage appears in Upgren (2005: 3). The laws are consulted to justify similar counterfactuals in Neil Comins, *What If the Moon Didn't Exist: Voyages to Earths That Might Have Been* (1993), cited by Maudlin (2007: 65). The same applies to books like *The Confederate States of America: What Might Have Been* (2005) by the historian Roger Ransom.

28. Defenders of such ideas include Bennett 1984, Carroll 1994, Chisholm 1946 and 1955, Goodman 1947 and 1983, Horwich 1987, Jackson 1977, Mackie 1962, Pollock 1976, and Strawson 1952. Seelau, Seelau, Wells et al. 1995 offer a psychological perspective on the way "that counterfactual thoughts are

restricted to those that are plausible given the natural laws operating in the world" (p. 66).

29. NP requires more defense than I can afford to give it in the main text. In this longish (OK, *very* long) endnote, I will look at three classes of apparent counterexamples to NP and explain why they all fail. (See also my 2000.) Here is an example of *Class #1:*

> Suppose that after work, two physicians discuss their day. The first says to the second: "So the nurse rushed over and reported that the patient had been accidentally injected with the syringe marked *A*. That syringe was intended for the lab; it was filled with arsenic—*A* for 'arsenic.' So I hurried over to the patient's bedside, although I knew, of course, that there was nothing I could do. I waited for the inevitable. But the most remarkable thing happened: the patient did not die. So our dismay turned to excitement: we thought we had a reportable case and prepared to write a stunning article for *The New England Journal of Medicine*. Then I checked the syringe. The nurse had misread the label; it turned out to be *H*, not *A*. So it contained no arsenic after all. Although the patient was out of danger, I must confess that we were all a bit disappointed. Had the syringe been filled with arsenic, we would have discovered that it is not a natural law that such a large dose of arsenic is lethal."

The truth of "Had the syringe been filled with arsenic, then it would not have been a law that such a large dose of arsenic is lethal, since the patient lived" appears to violate NP: the antecedent is logically consistent with the laws, but the consequent is not. However, I suggest that in this conversational context, the counterfactual's antecedent is implicitly "Had the syringe used to inject the patient been filled with arsenic and the patient lived," which is logically inconsistent with the laws, so the truth of this counterfactual conditional is no threat to NP.

Of course, I need some principled reason (beyond a wish to save NP!) for supposing that the antecedent tacitly includes "and the patient lived." Here is my reason. The second physician could respond to the first, "Don't feel too bad; you didn't come all that close to having made a great discovery. Had the syringe been filled with arsenic, would the patient still have lived?" The first physician should admit, "No, I suppose not. Had the syringe been filled with arsenic, the patient would have died." But how can the first physician believe this conditional while also believing the conditional she asserted earlier ("Had the syringe been filled with arsenic, then it would not have been a law that such a large dose of arsenic is lethal")? Because the earlier conditional's antecedent tacitly included "and the patient lived," unlike the later conditional's antecedent.

Suppose we say "$p \;\square\!\!\rightarrow q$ because r" (e.g., "Had the syringe been filled with arsenic, then it would not have been a law that such a large dose of arsenic is lethal, since the patient lived") and then ask "Is it true that $p \: \Box\!\!\rightarrow r$?" If the first counter-factual's antecedent p tacitly included r, then this question has the effect of stripping r out of p in any of p's appearances as an antecedent in the near future. (If r remains implicit in p, then the question "Is it true that $p \: \Box\!\!\rightarrow r$?" is trivial, and so to take p as implicitly including r is to give an uncharitable reading of the question.) Therefore, if the answer to the question "Is it true that $p \: \Box\!\!\rightarrow r$?" is "No," then r must have been preserved under the original counterfactual's antecedent only by virtue of its having been implicit in that antecedent. Contrast the arsenic example, where the answer is "No," with another example: "Had the match been struck, it would have lit, since the match was dry." Had the match been struck, would it still have been dry? Yes! Hence, that the match is dry does not have to have been implicit in the original counterfactual's antecedent in order for the match's dryness to be preserved under that antecedent. The match's dryness is preserved under the antecedent in virtue of (for lack of a readier metaphor) the "metric" determining the "closest possible world" where that antecedent obtains.

The same approach works well in other examples. Consider: "It would take a miracle for me to be on Jupiter within the next 10 seconds. Since I am starting from here on Earth, which is more than 10 light-seconds from Jupiter, I would have to violate the law prohibiting superluminal travel, were I to be on Jupiter sometime within the next 10 seconds." The truth of that counterfactual apparently violates NP: its antecedent appears to be logically consistent with the laws, whereas its consequent is not. But suppose that having asserted this counterfactual, I am then immediately asked, "Were you to be on Jupiter sometime within the next 10 seconds, would you be here now?" I should reply, "No. Were I to be on Jupiter sometime within the next 10 seconds, I would now have to be much nearer to Jupiter than I actually am." So the initial conditional's antecedent is implicitly "Were I to be on Jupiter sometime within the next 10 seconds starting from here, more than 10 light-seconds from Jupiter." Since this antecedent contradicts the laws, NP permits this conditional's truth. (I'll return to this below.)

On to *Class #2*. Here are three apparent violations of NP:

Gödel was a great logician. Had he denied the "law" of double negation, then the "law" of double negation might well have been false.

The experiment was designed to measure the mass of the electron. The dial giving the output of the experiment pointed to 72.9, which revealed the electron's mass to be 9.11×10^{-31} kg. But had the dial pointed to 50, then the electron's mass would have been only about two-thirds of 9.11×10^{-31} kg.

The half-life of Iodine-131 is 8.1 days. Had every one of the many atoms of ^{131}I in the history of the universe decayed before becoming 8.1 days old, then ^{131}I's half-life would almost certainly have been less than 8.1 days.

Someone infatuated with the context-sensitivity of counterfactuals might well suppose that in a context where Gödel's logical acumen is salient, the first of these counterfactuals is true. However, I disagree: Had Gödel denied the "law" of double negation, then he would have been a lousy logician. I believe the corresponding *indicative* conditional: If Gödel (of all people) believed the principle of double negation to be false, then it may well actually be false. Just as Gödel's opinion is a good indicator of the logical necessities, so likewise the reading on the dial is a good indicator of the electron's mass (and the decay of ^{131}I atoms is a good indicator of ^{131}I's half-life); if the dial reads 50, then the electron's mass is not 9.11×10^{-31} kg (an indicative conditional).

Let's fix firmly in mind the contrast between indicative and subjunctive conditionals:

Indicative: If Oswald did not shoot Kennedy, then someone else did. (True)

Subjunctive: If Oswald had not shot Kennedy, then someone else would have. (False)

Indicative: If the United States invaded Sweden last year, the event received tremendous media coverage. (False)

Subjunctive: Had the United States invaded Sweden last year, the event would have received tremendous media coverage. (True)

Just as our assertion of the Kennedy indicative reflects the way we believe we should have to revise our opinions upon learning that Oswald did not shoot Kennedy, so likewise our assertion of the Gödel indicative reflects the way we believe we should have to revise our opinions upon learning that Gödel denied the law of double negation.

That the above examples involve indicatives masquerading as counterfactuals is also suggested by the tension in "The principle of double negation is definitely a genuine law of logic, and had Gödel denied the principle, then it might well have been false." (The tension is absent from "The principle of double negation is definitely a genuine law of logic, and had Gödel denied the principle, then he would have been a lousy logician.") This tension is present in the following analogous remark:

Catullus in *The Marriage of Peleus and Thetis* (Catullus LXIV) definitely said "Emathiae tutamen opis carissime nato" (line 324), where "opis" is the genitive of Ops (the mother of Jupiter), so Catullus said, "Protector of Emathia, most dear to the son of Ops." But had every notable classicist read "opis" instead as "power," then *that's* what Catullus would have said, namely, "Bulwark of Emathian power, famed for thy son to be."

This sort of tension is often present in indicative conditionals: "The United States definitely did not invade Sweden last year" stands in marked tension with "If a U.S. invasion of Sweden was widely reported in the media last year, then the United States did invade Sweden last year." On the other hand, such tension is absent from counterfactuals: "The Yankees definitely won the game, and had Jeter dropped the ball, the Yankees would not have won the game." (And: "There was definitely no rain before the race, and had my horse won, there would have to have been rain before the race.")

We arrive at a counterfactual in *Class #3* by pretending the actual laws to be deterministic and the characters in *Pride and Prejudice* to be actual people. Suppose that Mr. Darcy and Elizabeth Bennett actually quarreled two days ago, and Elizabeth was then so cross that had Darcy asked Elizabeth for a favor yesterday, she would not have granted it. This counterfactual supposition seems to direct our attention to a "possible world" where Darcy asks Elizabeth for a favor but where until he does so, the events that transpire are the same as in the actual world. (For example, the quarrel still occurs.) For the counterfactual world and the actual world to coincide until the moment Darcy makes his request in the counterfactual world, but to diverge thereafter, requires that the actual (deterministic) laws be violated by Darcy's request; his action occurs without any of the causal antecedents required by the actual laws—"miraculously" (as Lewis puts it). Hence, had Darcy yesterday requested a favor from Elizabeth, the laws of nature would not have been the actual laws; there would have been different laws. This counterfactual conditional's truth violates NP.

(This appears to be one of Lewis's principal reasons for rejecting NP. The example comes from Downing 1959 and was popularized by Bennett 1974 and Lewis 1986b: 33–34. But if Lewis is correct, then why don't we ever in ordinary practice accept such counterfactuals as "Had Darcy requested a favor from Elizabeth, then the laws of nature would have been different"? For that matter, had Darcy requested a favor from Elizabeth and the actual natural laws failed to hold of the causal antecedents of Darcy's request, then they might well have failed to hold of later events as well. Although Lewis denies *that,* it is presumably true, just as had Coulomb's law been violated by uniformly charged spheres, then Coulomb's law might well have been violated by uniformly charged planes as well. If the laws would have been different, had Darcy requested a favor from Elizabeth, then those different laws might even have led to Elizabeth's granting Darcy's request. Here's another example [after Todd 1964: 104]. Had the ball been hit three feet higher, outfielder Jones would not have managed to catch the ball. But if the laws would have been different had the ball been hit three feet higher, then those different laws might have involved weakened gravity so as either to have enabled Jones to jump higher or to have led Jones, accustomed to the weakened gravity, to play deeper—allowing him to catch the ball.)

Faced with this apparent violation of NP, we might try various easy ways out. None of the following four succeeds:

(i) If the actual laws are indeterministic, then there is presumably a possible world governed by the actual laws where Darcy makes his request but the actual course of events is duplicated until just about that moment. (Perhaps some indeterministic, exceedingly unlikely event in Darcy's brain causes him to make his request.) However, it would be too astonishing for our longstanding counterfactual practice to have anticipated relatively recent scientific developments by implicitly presupposing that the natural laws are indeterministic. In other words, if the counterfactuals that we ordinarily accept presuppose indeterminism, then until very recently, no one was justified in believing in any of those counterfactuals (since no one was justified in believing in indeterminism)—which I take as showing that this approach fails.

(ii) I am also disinclined to say that if the laws are deterministic, then had Darcy requested a favor yesterday from Elizabeth, the state of the world at each prior moment would have to have been different in some respect from what it actually was. This counterfactual may be true in certain contexts (for instance, where we are illustrating what a remarkable doctrine determinism really is!), but it is not true in the context where we were originally considering what would have happened, had Darcy requested a favor from Elizabeth. In that context, we pay no attention at all to what the past would have to have been like (considering the actual laws) in order to allow Darcy to make his request. Regarding the view that the world's state at each prior moment would have to have been different from what it actually was, Bennett says: "Those remarks concern the exploration of determinism. To force them into our thinking about particular subjunctive conditionals... is to ride roughshod over a patent fact, namely, that when we think a subjunctive conditional we... do not dig into how [the antecedent] might have come about" (2003: 224–25). I agree with Bennett's thought. Indeed, I will take this thought more seriously than Bennett does; it contains the heart of my response to this problem for NP.

(iii) Perhaps the deterministic laws allow for a history that differs only negligibly from the actual world's at all times until Darcy makes his request (at which time, the differences explode). Perhaps. However, the truth-values of various counterfactuals are surely not hostage to whether this is so. That this is so would again amount to a sophisticated scientific discovery that our longstanding practice of counterfactual reasoning should not be credited with having anticipated. We can justly believe in the truth of various ordinary counterfactuals without having any reason to be confident that the deterministic laws permit a history differing only negligibly from the actual world's until Darcy's request.

(iv) It might be suggested that this example belongs to Class #1: When we are told "Had Darcy asked Elizabeth for a favor yesterday, she would not have granted it, because of their earlier quarrel," we reply by asking, "But had Darcy asked Elizabeth for a favor yesterday, would they have quarreled earlier (and would the rest of history until yesterday have been as it actually was)?" No! So by my earlier argument, the first counterfactual's antecedent was really "Had Darcy asked Elizabeth for a favor yesterday despite their earlier quarrel (and the rest of actual history until yesterday)," which is logically inconsistent with the actual (deterministic) laws, making this counterfactual's truth no threat to NP.

I reject this way out, too. The question "Had Darcy asked Elizabeth for a favor yesterday, would they have quarreled earlier?" demands that we turn our attention to what the past would have to have been like in order for Darcy to have made his request—which (as Bennett rightly remarked) was utterly irrelevant in the context where we originally considered what would have happened, had Darcy asked Elizabeth for a favor yesterday. My test for uncovering clauses implicit in the counterfactual antecedent works only if the question asked in the test ("Is it true that $p \,\square\!\!\rightarrow r$?") leaves us in the same context of interests and concerns as before. If there is a shift in the context as a result of that question's being asked, then there is a change in the metric determining the "closest possible world" where the antecedent obtains, and so the test fails to separate what was preserved under the original counterfactual antecedent by virtue of having been implicit in that antecedent from what was preserved by virtue of the original closeness metric. The test question "Had Darcy asked Elizabeth for a favor yesterday, would there have been a prior quarrel?" almost inevitably turns our attention from Elizabeth's anger to Darcy's pride, which was not a consideration in the original context. (Darcy is a proud man; he would not have asked for a favor unless he had good reason to expect it to be granted.) Compare a case in which the test is properly deployed: Both the original counterfactual ("Were I to be on Jupiter sometime within the next 10 seconds, I would have to violate the law prohibiting superluminal travel") and the final counterfactual in the test ("Were I to be on Jupiter sometime within the next 10 seconds, I would now have to be much nearer to Jupiter than I actually am") are making the same point: that I am now too far from Jupiter to be able to get there within the next 10 seconds. These two counterfactuals are being entertained in the same context, unlike in the Darcy-Elizabeth case. (If we try resolutely to hold the context fixed when asking "Had Darcy asked Elizabeth for a favor yesterday, would there still have been a prior quarrel?" we must answer, "Yes: that's why Elizabeth would have refused to grant Darcy's request.")

To deal with this kind of apparent counterexample to NP, I suggest that talk of "possible worlds" is in certain respects misleading even as a metaphor for the truth-conditions of counterfactuals. (Elsewhere in this book, I mention two other reasons for regarding possible-worlds talk as inapt: I deny Centering [see chapter 3, notes 61 and 64, and chapter 4, note 1] and [in chapter 2] I suggest that some counterfactual conditionals with logically impossible antecedents are nontrivially true in some contexts.) In the familiar sort of context where it is true that Elizabeth would not have granted Darcy's request, had Darcy asked her for a favor yesterday, the events leading Darcy to make his request are off-stage, behind the scenes, out of sight—not part of the counterfactual "world" at all. That there would have been a "miracle" (a violation of the actual laws) demands the truth of such counterfactuals as "Had Darcy asked Elizabeth for a favor, then the laws of nature would have been different" or "...then Darcy would have to have either forgotten about the quarrel or believed that Elizabeth would have forgotten it." But no such counterfactual is true in this context. Apart from such salient details as the quarrel (and such general background facts as Darcy's acquaintance with Elizabeth and the mores of early-nineteenth-century English society), no events prior to yesterday are in play in this context. It is not the case that they would still have occurred, had Darcy requested a favor from Elizabeth. Rather, in this context, no counterfactual conditional regarding those events is true and none is false.

Of course, a remark that is irrelevant to a given conversation may violate conversational norms or waste everyone's time, but it is not ordinarily thereby deprived of truth-value! However, a counterfactual conditional acquires a truth-value only if the context helps to supply it with one. If the conditional concerns what is "offstage," then the context fails to make the necessary contribution, so the conditional lacks truth-value. The function (see note 15) mapping the counterfactual sentence to different propositions, depending on the context, fails to map the sentence to any proposition at all in a context where the counterfactual concerns offstage matters. Therefore, the conditional is not even false. It falls into a truth-value gap.

Instead of taking a counterfactual antecedent as directing our attention to an entire possible world-history, I see it as generating something more like a short story—a brief piece of historical fiction. (Kim and Maslen [2006] also make this comparison.) A story typically begins "Once upon a time, p obtained" and the default is that it is irrelevant how p managed to come about in the first place. (That default may be overridden; some of p's antecedents may later be filled in and play a special role in the story.) If someone interrupts the start of the tale to ask "How did p come to pass?" the storyteller replies, "Never mind. That doesn't matter. That's not part of the story. Please let me get on with it." Similarly, in the familiar sort of context in which we were considering whether

Elizabeth would have granted Darcy's request, the counterfactual supposition "Had Darcy requested a favor from Elizabeth" is entertained without regard to how this supposition managed to obtain.

In a Sherlock Holmes story, for instance, although the laws of nature are the actual laws, and the story together with those laws may entail something about the events a billion years before, nothing about what happened then is part of the story. The story begins long after that time. It is neither true nor false in the story that various events transpired a billion years before Queen Victoria's reign. (My claim is not that a reader, wondering about such events, should conclude that they are left open by the story. Rather, my claim is that the reader does not wonder about them at all.)

Similarly, the "miracle" allegedly needed for Darcy to have asked Elizabeth for a favor yesterday is not part of the short story evoked by the counterfactual supposition; no "miracle" is required to attach Darcy's request onto actual history because that history (except for certain salient details and certain general background facts) is not part of the story. (Although $(p \:\square\!\!\rightarrow q)$ and (q logically entails r) seems to demand $(p \:\square\!\!\rightarrow r)$, I do not believe that this principle holds generally, since in the context where p is being entertained, whether or not r would have held may be "offstage." Whatever we need this principle to do can be done by its restriction to cases where r is not offstage.)

A short story does not say what happens after "The End." By the same token, in the context where we consider whether Elizabeth would have granted Darcy's request, we do not consider whether Darcy would later have wondered what had come over him, leading him to make such a request. Yet if the counterfactual supposition evokes an entire possible world-history, then we surely would have to accept such counterfactuals as "Had Darcy requested a favor from Elizabeth, then after she rebuffed him, he would have been puzzled as to how he had come to make such a request in the first place, as well as worried about whether he would again engage in such erratic, uncharacteristic behavior."

Even small children understand how stories work. A fact does not have to be mentioned explicitly by the story's text for it to be "in the story." Even if Sherlock Holmes never explicitly says why he decided to interview the groundskeeper rather than Queen Victoria, a reader understands his reason: because the former might have seen or heard something on the fatal night, whereas the latter was nowhere near the scene of the crime and therefore (considering the laws of nature) cannot assist Holmes in his inquiries. Neither Holmes's reasoning nor Queen Victoria nor the natural laws are explicitly mentioned by the text, yet they are part of the story. Within the story, everything proceeds in accordance with the actual laws of nature because they are the laws in the story. (*The Adventures of Sherlock Holmes* is a work of historical fiction, not *science* fiction.)

Just as small children understand how stories work, so they also understand how counterfactuals work. Upon leaving my home one morning, I closed the front door, but when I went to lock it (my door does not lock automatically), I realized that I had mistakenly left my keys inside. My nine-year-old son Abe remarked, "It's lucky that the door is unlocked. If it had been locked, you wouldn't have been able to get your keys." I replied, "If it had been locked, *I* would have locked it, and so I would already have had my keys." Abe recognized that he and I were not disagreeing. Abe's counterfactual invoked a story that began, "Once upon a time, we were outside (just as we actually were—without our keys), but with the door locked." How that initial condition arose is not part of this story, whereas my counterfactual concerns exactly that.

Bennett (2003: 284–85) says that some counterfactuals (forming a "mildly degenerate though quite common kind") involve "no thought about a possible history for the antecedent," as when "Charles's wife remarks sarcastically, 'If Charles had been CEO of Enron, the accounting fraud would not have lasted a week,' because Charles is incompetent with money." Bennett says, "What it omits, which a Lewis-type grounding includes, is a thought about whether Charles is a financial incompetent not only [in the actual world] but also at the closest worlds at which he runs Enron.... [She regards] as an irrelevant nuisance the question of whether Charles could have *come* to run Enron while still financially incompetent." I agree with Bennett about this case—except I regard it as perfectly typical rather than as belonging to "a marginal and uninteresting sort that I cheerfully relinquish to any philosopher who wants to spend time on them" (p. 255).

I am not trying to suggest that counterfactual conditionals have truth-conditions in terms of short stories. Furthermore, despite its potential for misleading us, I will occasionally indulge in talk of "possible worlds" in connection with counterfactual conditionals. But we must proceed cautiously.

30. For example: Bennett 2003: 224; Foster 2004: 90–91; Pollock 1974: 201; Stalnaker 1984: 155–56; Swartz 1985: 53–54; Van Inwagen 1979: 449–50.

31. In my definition of "sub-nomic stability," I have required that any sub-nomically stable set be logically closed (as far as sub-nomic claims are concerned). I have imposed this requirement solely for the sake of simplicity. Without this requirement, there would be many sub-nomically stable sets containing some but not all of the first-order laws, each having Λ as its logical closure (sub-nomically speaking). It is easier to have only Λ among the stable sets than to have all of these others joining it. However, I could have omitted logical closure from the requirements for stability and reformulated my later remarks accordingly. The sets that would have qualified as "sub-nomically

stable" under that definition are exactly those having (sub-nomic) logical clo-sures that are sub-nomically stable under my definition.

32. See chapter 3, notes 61 and 64, and chapter 4, note 1.

33. Suppose that *g* concerns a racecar driven by Smith. Take a context where Smith's deep commitment to racing at top speed is salient. He would have done whatever it took to race faster. Had *g* been false or Jones worn an orange shirt, then *g* would have been false; Smith would have had something done to the car, not to Jones's shirt, since the latter could not have increased the car's speed. In contrast, take a context where Smith's belief that he has fine-tuned his car into peak operating condition is salient. Had *g* been false or Jones worn an orange shirt, then Jones would have worn an orange shirt; Smith would not have allowed anything to mess with his car.

34. Of course, philosophers who deny NP deny that it is more outlandish. My argument here will not persuade them to adopt NP. My purpose in offer-ing this argument is instead to explain why it is not unprincipled to acknowl-edge context's tremendous influence on what is preserved under a given counterfactual supposition, and yet to insist that in any context, the laws are preserved under any sub-nomic supposition logically consistent with them. (That explanation begins in the next sentence.)

35. The step from $(\sim s \text{ or } \sim t) \,\Box\!\!\rightarrow (t \,\&\, (\sim s \text{ or } \sim t))$ to $(\sim s \text{ or } \sim t) \,\Box\!\!\rightarrow \sim s$ seems to use a principle that I rejected near the end of note 29: if $(p \,\Box\!\!\rightarrow q)$ and $(q$ logically entails $r)$, then $(p \,\Box\!\!\rightarrow r)$. But it suffices to use the principle with the added requirement that r not be "offstage." I presume that in at least one con-text where the counterfactual supposition $(\sim s \text{ or } \sim t)$ is entertained, neither s nor t is offstage.

36. If Γ contains all of the broadly logical sub-nomic truths (which include the [narrowly] logical, conceptual, mathematical, and metaphysical sub-nomic truths) and Σ does, too, then the *reductio* could have been accomplished a bit differently. As before, Γ's stability entails $(\sim s \text{ or } \sim t) \,\Box\!\!\rightarrow \sim s$. Since Σ is stable, its members would all still have held, had $(\sim s \text{ or } \sim t)$, and so in particular, s would still have held. Thus, $(\sim s \text{ or } \sim t) \,\Box\!\!\rightarrow (s \,\&\, \sim s)$. Now $(\sim s \text{ or } \sim t)$ is logically consis-tent with the broadly logical sub-nomic truths (since $(\sim s \text{ or } \sim t)$ is logically consistent with Σ and Σ contains the broadly logical sub-nomic truths). There-fore, $(\sim s \text{ or } \sim t) \,\Box\!\!\rightarrow (s \,\&\, \sim s)$ conflicts with the stability of the broadly logical sub-nomic truths, giving us our *reductio*. In other words, for this conditional to be true would be for a logical impossibility to obtain under a broadly logical possibility. But anything that would have happened, had something broadly logically possible happened, must also qualify as broadly logically possible. (The same may not be the case for narrowly logical possibility: perhaps in some contexts, it is true that had there been a round square (a broadly logical

impossibility but a narrowly logical possibility), then (s & $\sim s$) would have held for some s. This point will arise again in chapter 2.)

We might have tried a slightly different *reductio:* For Γ and Σ both to be stable, ($\sim s$ or $\sim t$) $\Box\!\!\rightarrow$ (s & $\sim s$) must hold in all contexts. If ($\sim s$ or $\sim t$) is broadly logically possible, then this conditional is false in all contexts (because the broadly logical truths are stable). If ($\sim s$ or $\sim t$) is broadly logically impossible (but narrowly logically possible, of course, since it is logically consistent with Γ), then even if there are some contexts where ($\sim s$ or $\sim t$) $\Box\!\!\rightarrow$ (s & $\sim s$) holds, it presumably fails in some contexts.

37. Airy 1830.

38. Ehrenfest 1917.

39. If we take Newton's second law of motion to concern not the acceleration associated with the net force, but rather the contribution to the acceleration associated with a given force (whether net or component), then the law of the composition of forces follows from the second law. That is how Newton presents the parallelogram of forces (as a "corollary"), an approach followed by Thomson and Tait 1888: 244 (secs. 255–56). Others (such as Poisson) regard the parallelogram of forces as transcending the fundamental dynamical law.

40. Wigner 1972: 13.

41. Einstein 1935: 223.

42. For references to much of the early literature, see Berzi and Gorini (1969: 1518); for more recent references, see Pal (2003).

43. Penrose 1987: 21.

44. For example, Lévy-Leblond 1976: 271.

Chapter 2

1. Alias "nomic," "nomological," and "physical" necessity.

2. My aim is not to determine which system of logic (e.g., classical, free, or paraconsistent) accurately specifies the logical truths, but rather to understand what logical necessity is, whatever the logical truths turn out to be.

3. Alias "analytical" and "analytic" necessity.

4. Thomson (1990: 18) offers this "melodramatic" example. Of course, some normative truths (such as "Lange's behavior in the—affair was less than morally impeccable") are not morally necessary since they depend on morally contingent facts (concerning, e.g., Lange's behavior). But the "moral laws" (i.e., the theorems of the correct ethical theory) possess moral necessity.

5. I am concerned only with the modality of facts—that is, with *de dicto* modality, not with *de re* modality. Moreover, I am concerned only with

modalities where all necessities are truths, i.e., where nothing impossible happens. I thereby set aside doxastic, bouletic, deontic, and teleological modalities, since (for example) that all persons are legally obligated to obey the nation's laws does not entail that all do.

I set aside epistemic modalities as well. Although all epistemic necessities are truths, this is so on a technicality: that knowing p requires p. Epistemic modality is better understood as a species of doxastic modality.

Teasing occurs despite the moral necessity that no child be teased. Nevertheless, all moral necessities are truths: that it is morally necessary that all teasing is wrong entails (indeed, explains why it is) that every actual case of teasing is wrong.

6. Lewis 1973: 72–77; 1983; 1986b; 1999a.

7. Ellis 1999, 2001, and 2002.

8. Leibniz 1902: 4 (II).

9. However (setting aside in this note our convention to reserve lower-case italicized English letters for sub-nomic claims), if we take p to be "It is [not] a law that m" and p is true, then p is obviously naturally necessary on the latter definition but is not obviously so on the former. On Lewis's "Best System Account" of law (which I shall discuss momentarily), the latter definition would incorrectly deem the following naturally necessary: The world's history does *not* consist solely of the uniform motion of a single electron alone in the universe forever. That is because on Lewis's view, the actual laws could not be the laws of such an impoverished universe; they would not all belong to the Best System there. Hence, Lewis adopts the former definition of natural necessity.

10. From the *Star Trek* episode "The Naked Time." My thanks to John Roberts for suggesting this example.

11. Perhaps Lewis restricts the scope to possible worlds containing no instances of "alien properties." If so, then (unlike many philosophers who take the laws as supervening) Lewis does not regard the laws' supervenience as metaphysically necessary. However, perhaps Lewis so restricts his general thesis that everything supervenes on the Humean base, while intending his account of laws to be necessary.

12. Lewis's elite properties are not exactly those ascribed by my "sub-nomic" claims. For example, sub-nomic claims ascribe chances and perhaps dispositions, too (see chapter 1, notes 23 and 25).

13. To accommodate quantum-mechanical "entanglement," this pointlike requirement might be relaxed.

14. Lewis specifies that for a system that includes p to be eligible for the competition, it must also include the claim that p never had any chance of not being true. I will discuss this requirement in chapter 3.

15. If nature is "unkind" to us in that which system qualifies as "best" is not robust in this way, then "there would be no very good deservers of the name

of laws. But what of it? We haven't the slightest reason to think the case really arises" (Lewis 1999: 233).

16. However, Lewis's account presupposes that only certain properties qualify as perfectly "natural."

17. "The chess-board is the world, the pieces are the phenomena of the universe, the rules of the game are what we call the laws of Nature.... Education is learning the rules of this mighty game. In other words, education is the instruction of the intellect in the laws of Nature..." (Huxley 1871: 31–32). For similar remarks, see Feynman 1967: 36, 59. The rules of Conway's "game of life" (and other cellular automata) are frequently termed its "laws" (e.g., Gardner 1970). I'll return to this metaphor in chapter 4, section 4.8.

18. Davies 1995: 256; Dorato 2005.

19. Of course, if a player fails to castle under circumstances in which castling would have been advisable, then this fact may constitute good evidence that castling was not permitted. But it is not decisive evidence: the player might be unskilled or careless. The moves actually made fail to *determine* whether castling was permitted (unless the pieces involved in the move are individuated by the moves they could make—the analogue of "scientific essentialism," which I discuss later).

20. This example is from Earman (1986: 100); I discussed it in my 2000.

21. For more sophisticated arguments against the laws' supervenience on the Humean mosaic, see Carroll (1994: 60–68) and Tooley (1977: 669–72).

Counterfactuals positing radically impoverished universes figure in scientific practice—for example, when scientists investigate whether a complex model is at all promising. Most of the model's complexity can be ignored when it is applied to a radically impoverished case. Whether the model yields the correct answer for such a case is easy to see.

If some lone-electron world has laws covering all sorts of particles unrepresented there, mightn't the actual world also have oodles of laws covering kinds of particles that go forever uninstantiated? Perhaps; for instance, particles that would be created under extreme conditions that happen never to be realized. However, the "closure laws" (such as that all charged leptons are muons or electrons or taus) prevent an unlovely proliferation of laws and kinds. (See chapter 3.)

22. Lewis rejects this principle, for reasons I discuss in chapter 1, note 29.

23. This counterfactual's truth allows there to be some "possible worlds" containing many electrons where the inter-electronic force violates Coulomb's law and accords with some non-Coulombic law instead. But none of these worlds is the *closest* many-electron world to the lone-electron world closest to the actual world.

24. I will return to this argument in chapter 4, section 4.2.

25. Armstrong 1978, 1983, 1997. Similar views appear in Dretske 1977 and Tooley 1977, 1987.

26. However, Armstrong's "naturalism" demands that all universals be instantiated. (Tooley permits uninstantiated universals.) Armstrong develops further apparatus to account for uninstantiated laws.

27. For Armstrong, laws are facts such as that Fness nomically necessitates Gness; what is naturally necessary are their consequences, such as the fact that all F's are G. For Lewis, laws are facts (such as that all F's are G) figuring in the Best System, and they are naturally necessary. I am fudging this difference in order to highlight how both views are vulnerable to the same kind of objection. Of course, scientific essentialism is invulnerable to this objection, since on that view, laws really are necessary—*metaphysically* necessary. (That's *too* necessary, I believe.)

28. Lewis 1986b: xii; cf. Lewis 1983: 366; Van Fraassen 1989: 98.

29. Van Fraassen 1989: 47.

30. Lewis 1999: 232.

31. "Posit all the primitive unHumean whatnots you like. (I only ask that your alleged truths should supervene on being.) But play fair in naming your whatnots. Don't call any alleged feature of reality 'chance' *unless you've already shown* that you have something, knowledge of which could constrain rational credence" (Lewis 1999a: 239, my emphasis; see also 1986b: xv–xvi).

32. An account of lawhood as ontologically primitive (see Maudlin 2007) is likewise vulnerable to this recipe for causing trouble: an account that fails to unpack lawhood has few resources with which to explain how laws manage to qualify as necessary—as possessing a distinct species of the same genus as other varieties of genuine necessity.

33. Lewis 1986b: 77; see also Lewis 1986a: 8.

34. Such views are common (e.g., Lewis 1973: 7; Divers 2002: 5; Von Fintel 2006: 21–23) and can take sophisticated forms (Kratzer 1991). Here *p* must be a claim that is restricted to describing a single world, not a claim like "There exists a plurality of worlds."

35. Jockl from Kratzer 1991: 640; Pavarotti from Rocci 2005: 231; appointment from Divers 2002: 4; prunes from Wertheimer 1972: 93. Lycan (1994) offers a cornucopia of further examples.

36. Objection: It is not true that *p*'s natural necessity explains why *p* obtains. Rather, to respond to "Why *p*?" with "It is a law that *p; p* couldn't have been otherwise" is to say that *p* has no explanation and needs none. That *p* is a law excuses *p* from needing an explanation.

I don't think I have to disagree. If *p*'s genuine necessity (e.g., its natural necessity) explains *p*'s truth, whereas *p*'s merely conversational necessity does not, then genuine necessity carries special metaphysical weight. If, on the other hand, p's genuine necessity excuses p from having to have an explanation (on the grounds that p is inevitable) whereas p's merely conversational necessity does not, then once again, genuine necessity carries special metaphysical weight. (Thanks to John Roberts here.)

Objection: If p's natural necessity explains why p obtains, then it is too easy to explain p. One merely has to say "It is necessary that p." Reply: (i) Not every p is naturally necessary, so we cannot answer every why-question this way. (ii) Even if p is naturally necessary, we sometimes require an explanation appealing to more fundamental laws.

An explanation of the form "Why p? Because p couldn't have been otherwise; it is a law that p" is no less explanatory than an explanation of some accident that invokes initial conditions as well as laws. It just manages to do without the initial conditions.

37. A definition of "logical truth" given by Quine 1961: 22–23.

38. A definition of "logical truth" given by Quine 1970: 50.

39. Tarski 1956: 418–20.

40. To say that the fact expressed in a given context by "It is necessary that q" involves a "merely conversational" or "merely relative" necessity, rather than some variety of "genuine" necessity, is *not* to say that q's necessity is not objective. That q follows logically from a certain (tacitly designated) set of facts is a perfectly objective matter.

41. Stalnaker (1968); Williamson (2005).

42. Augustine (1982: 112). Six is a "perfect" number because it is the sum of its divisors (1, 2, and 3) excepting itself.

43. And had something possible happened, then whatever would have happened, had something possible happened, is also possible (and so forth). For simplicity, I will set these nested counterfactuals aside for now.

44. To keep things relatively uncluttered, I typically omit quotation marks (and corner quotes) around symbolic expressions, as long as there is little risk of confusion.

45. Cruttwell (1991: 266); I am following Bulhof (1999) in using this example.

46. Mill 1874: III, 5, vi.

47. This notion arose in one version of my argument that the sub-nomically stable sets are nested (see chapter 1, note 36) and also in my argument that Λ is sub-nomically stable.

48. An extreme example involves what some philosophers (e.g., Lewis 1986a: 7) call "historical necessity": what's true in every world that matches ours perfectly up to now (and also includes the actual world's logical, conceptual, metaphysical, mathematical, and natural necessities, I presume). If the universe is deterministic, then all facts (at least, all physical facts and whatever supervenes upon them) turn out to possess historical necessity.

49. Of course, we could *invent* a relative modality. For example, we could stipulate that *p* is "Washington-necessary" exactly when *p* follows logically from the fact that George Washington was the first president of the United States. Then obviously, not every fact is Washington-necessary. But this "variety of modality" does not arise from some natural-language conversational context, so it is irrelevant to my project, which was to take the modalities figuring in natural language and to distinguish those that are genuine from those that are merely relative. For a modality that fails to arise in some ordinary conversational context, there is no tacit sense of what is conversationally relevant when it is salient, and so my conjecture that all of the conversationally relevant facts qualify (by M) as necessary implies nothing.

50. Several species of merely relative modality can seemingly be in play in the same breath rather than at different stages in the conversation. The minutes of the White House meeting might include the remark: "The president was informed that although it is technologically possible to manufacture 100 million doses of avian flu vaccine by next year, the limits of current production capacity makes it impossible to do so." On the other hand, this remark might be construed as involving not two species of merely relative modality, but rather a single species in terms of which it is possible for factories using current technology to manufacture 100 million doses of avian flu vaccine by next year, but impossible for the factories currently extant to do so. (See section 2.13.)

51. See note 43 above and chapter 1, section 1.8.

52. In this chapter, I consider only sub-nomic truths. In chapter 3, I argue that meta-laws, which are not sub-nomic, form *nomically* stable sets.

53. Of course, it belongs to the set of *all* sub-nomic truths, and if Centering holds (as I mentioned in chapter 1), then that set is sub-nomically stable. But (even setting aside my claim that Centering does not hold) this set's "maximal resilience" fails to give it a corresponding species of necessity, since its sub-nomic stability arises from the fact that there are *no* sub-nomic counterfactual suppositions with which it is logically consistent. We could, I suppose, take the set of all sub-nomic truths as corresponding to the zeroth grade of necessity, the degenerate case. (We could likewise weaken the notion of "sub-nomic stability" so that the null set possesses sub-nomic stability, though again of a trivial sort.)

54. Or perhaps, although every member would still have held, there is some member *m* where ~*m* might also have held, under such a counterfactual supposition *p*. Moreover, if we bear in mind the nested counterfactuals required for sub-nomic stability, then there need not be any such counterfactual supposition *p*. But then there must be a possibility *p* under which, had some other possibility *q* obtained, some necessity *m* might not still have held, violating "Had something possible happened, then whatever would have happened, had

something else possible happened, must also qualify as possible"—or some further requirement for stability (involving more nested counterfactuals) must be violated. For the sake of simplicity, I ignore these cases; the same argument can be given for them.

55. If the narrowly logical truths form the smallest sub-nomically stable set, then it might seem circular to elaborate the kind of necessity they possess in terms of sub-nomic stability, since (unlike the natural laws), the narrowly logical truths play a special role in the definition of "sub-nomic stability." For one thing, that definition demands that a sub-nomically stable set contain every sub-nomic *logical* consequence of its members. For another thing, the definition demands that a sub-nomically stable set be preserved under every sub-nomic counterfactual supposition with which the set is *logically* consistent. These references to logical truth do not *trivialize* an account of logical necessity in terms of sub-nomic stability. (It is not like saying "*p* is logically true if and only if ~*p* is logically false.") But these references would preclude our saying (in response to the Euthyphro question regarding logical necessity) that a sub-nomic truth's "logical necessity" is just its membership in the smallest sub-nomically stable set, since "sub-nomic stability" itself presupposes the distinction between the logical necessities and the contingent truths.

However, the threatened circularity can be avoided if "sub-nomic stability" is redefined as follows:

> Consider a nonempty set Γ of sub-nomic truths. Γ possesses *sub-nomic stability* if and only if for each member *m* of Γ (and in every conversational context),
>
> $\sim (p \diamond\!\!\rightarrow \sim\!m)$,
> $\sim (q \diamond\!\!\rightarrow (p \diamond\!\!\rightarrow \sim\!m))$,
> $\sim (r \diamond\!\!\rightarrow (q \diamond\!\!\rightarrow (p \diamond\!\!\rightarrow \sim\!m))), \ldots$
>
> for any sub-nomic claims $p, q, r, \ldots$ where ~*p* is not a member of Γ, ~*q* is not a member of Γ, ~*r* is not a member of Γ, ...

This definition avoids any reference to "logical consistency."

I believe that Γ is sub-nomically stable, according to this new definition, if and only if Γ is sub-nomically stable according to the original definition. (For details, see my 2005.)

56. Apparently suggested by Field 1989: 236–38.

57. Lewis (1986a: 154–55) gives the "axiom of unique charge" as neither obviously a metaphysical truth nor obviously a mere matter of natural law.

58. This point arose earlier in chapter 1, note 36.

59. That there is a stratum of natural law without the force laws, but containing the fundamental dynamical law and its mates, is part of what motivated

unease with the equality of inertial and gravitational masses in classical mechanics. Their equality (more precisely: their standing in the same ratio for all bodies—which, with suitable choice of units, can be made equal to 1) connects a force law (the gravitational one) to the fundamental dynamical law—a connection *between* the strata, threatening to disrupt the stability of the higher stratum: had the gravitational force law been different, would the fundamental dynamical law have been different to compensate, preserving the fact that the gravitational acceleration of falling bodies is independent of their masses? In terms of the stronger variety of necessity possessed by the fundamental dynamical law, the gravitational force law (and hence the equality of inertial and gravitational masses) is an accident. ("The equality of the two masses...was quite accidental from the point of view of classical mechanics" (Einstein and Infeld 1951: 227); "It is an accidental gift of nature" (Born 1961: 313).) But the equality of inertial and gravitational masses seems suspiciously nonaccidental—and not merely by hindsight assisted by general relativity. Hertz wrote (about 1884):"This correspondence must mean more than being just a miracle" (Blaser 2001: 2395).

60. Searle 1897: 341; cf. Hunt 1991: 91. Searle was an associate of J. J. Thomson at the Cavendish Laboratory in Cambridge.

61. In an undated letter (probably from early 1889) quoted in Nahin (1987: 126). Today FitzGerald is often remembered for his "contraction hypothesis" advanced to deal with the null result of the Michelson-Morley experiment.

62. That was Heaviside's view. He tried to derive an equation for the superluminal case.

63. Here is an alternative reconstruction of the argument that apparently involves no counterfactuals: It is a law that all forces have real-valued (rather than imaginary-valued) magnitudes (since $F = ma$, and accelerations and masses are real-valued), so it follows from Heaviside's law that a charged body's speed cannot exceed c—at least if it has been moving uniformly for a long time in the presence of another charged body.

But how does this argument arrive at the more general moral drawn by Searle and FitzGerald: that no charged body can move superluminally? The argument might continue: If the laws prohibit a charged body's speed from exceeding c when it is moving uniformly for a long time in the presence of another charged body, then presumably this is not a special case but applies even to nonuniform motion, and even in the absence of another charged body. I suggest that this thought ought to be cashed out as follows. Had there been a charged body moving superluminally but nonuniformly or in the absence of another charged body, then there might have been a charged body in uniform superluminal motion in the presence of another charged body. In other words: Let m be that no charged body's speed exceeds c when it has been moving uniformly for a long time in the presence of another charged body. (That m is a law is the narrower conclusion that followed above from Heaviside's law.) The following is false: Had there been a charged body moving superluminally but nonuniformly or in the absence of another charged body, then *m* would still have held. The laws' stability is thereby threatened. So for *m* to be a law, the laws must prohibit charged bodies from any superluminal motion, not merely from uniform superluminal motion in the presence of another charged body.

Thus, by looking at the counterfactuals, we capture how the evidence for Heaviside's law counts also as evidence for a law prohibiting charged bodies from ever moving superluminally. The extent to which scientists were prepared to generalize the prohibition against charged bodies engaging in uniform superluminal motion in the presence of another charged body corresponds to the counterfactuals that scientists were prepared to accept specifying conditions under which there might have been a charged body in uniform superluminal motion in the presence of another charged body. (Interestingly, Searle, FitzGerald, et al. limited the prohibition to *charged* bodies. They seem not to have held that had an uncharged body moved superluminally, then some charged body might have done so too. Counterfactuals thus help to reveal the limits of their argument.)

64. Ellis 2001: 275; for similar remarks, see Bigelow, Ellis, and Lierse 1992; Ellis 2001: 205.

65. Ellis 2001: 276.

66. Ellis 2001: 278.

67. Chapter 1, note 36, suggests other ways in which this *reductio* might have been achieved.

Chapter 3

1. Within a single sentence, Spinoza (1951: 83) applied all three of these adjectives to the laws; Darwin (1868: 249) managed two. Descartes (2000: 28–29) addressed the laws' fixity in a letter to Mersenne (April 15, 1630): "God...established the laws of nature, as a King establishes laws in his kingdom.... You will be told that if God has established these truths, he could also change them as a King changes his laws. To which it must be replied: yes, if his will can change. But I understand them as eternal and immutable. And I judge the same of God." Cf. Earman 1989: 47.

2. Linde 1994: 48, 55.

3. Weinberg 1977: 143.

4. Our *beliefs* about the laws change (and so we change what we *call* "laws"). But these changes fail to show that laws change.

It is sometimes argued that there are laws of "special sciences" and that these laws were not laws until after their special subject-matter arose. For example, "The idea that Ohm's law has a timeless, transcendent existence, and has been 'out there,' lying in wait, for aeons until somebody built an electric circuit is surely ludicrous" (Davies 1995: 258). An analogous argument might be made regarding any putative law of biology, automobile repair, earth science, etc. Though I find this argument extremely dubious, I shall confine myself to whether the laws of fundamental physics can change.

5. "The hierarchy of laws has evolved together with the evolution of the universe. The newly created laws did not exist at the beginning as laws but only as possibilities" (Thirring 1995: 132; cf. Nambu 1985: 108–9; Stöltzner 1995: 50). Maudlin (2007: 12) countenances laws differing in different spatiotemporal regions (with no eternal, universal law regulating them).

6. Poincaré 1913/1963: 12–13. Admittedly, Poincaré regarded the laws discovered by science as not wholly mind-independent features of the world. Nevertheless, his argument for their immutability makes no appeal to any neo-Kantian views. Indeed, Shoemaker (1998: 75, n. 8) makes essentially the same argument.

7. Strictly speaking, Armstrong's account leaves no room for uninstantiated laws (since a universal, according to Armstrong, must be instantiated). But Armstrong's account allows for functional laws with uninstantiated values of the determinables. It construes functional laws as relations among second-order universals, such as the property of being a property involving having some value of electric charge, where this second-order universal is instantiated even if certain values of electric charge are not (Armstrong 1983: 113). So Armstrong's account would presumably allow for laws like (2) and (3) to be uninstantiated.

8. I could have argued instead that if (2) is a law in a given period, then since (2) must belong to a set that is stable for that period, the subjunctive conditional "Were squares four-sided, then (2)" is true, and so (since squares actually *are* four-sided) (2) is true—not merely true of that period. But (unlike the argument that I just gave in the main text) this argument fails to show that (2) is a *law* forever, though it does preclude (3)'s being an instantiated law during some period.

9. Here I appeal not to "Centering" (If q is true and m is true, then $q \,\square\!\!\rightarrow m$ is true), which I disavow, but to the following: If q is true and $q \,\square\!\!\rightarrow m$ is true, then m is true.

10. What if (3)'s lawhood is not *predetermined* to engage when the universe's age exceeds 10^{-10} seconds, but results from an indeterministic process? For example, suppose it is a law that when the universe is exactly 10^{-10} seconds old, there is a 50% chance that (3) will thenceforth be a law and a 50% chance that (2) will thenceforth be a law. (The statistical law we have just posited would be a meta-law: a law governing other laws. I will shortly say more about

meta-laws.) If, by chance, (3) turns out thenceforth to be a law, then it will apparently be a temporary law; before the universe is 10^{-10} seconds old, it is not a law that (3) holds after the universe is 10^{-10} seconds old, since before the universe is 10^{-10} seconds old, there is some chance that (2) rather than (3) holds afterward. (I will shortly say more about the laws' relation to chances.)

However, (3) cannot achieve temporary lawhood by this route if its temporary lawhood would require its membership in a set Γ that is stable (in the relaxed sense) for the period after the universe's age exceeds 10^{-10} seconds. Suppose that falsehood q exclusively concerns the period after the universe is 10^{-10} seconds old, and although q is logically consistent with (3) (and indeed with Γ), q's chance (at a given early moment) is much greater if (2) is true of the given period than if (3) is true of that period. Then (at least in certain contexts) had q obtained, the indeterministic process might well have had a different outcome and so (3) might not have been true of the given period. Therefore, (3) does not belong to a set that is stable (in the relaxed sense) for the period after the universe is 10^{-10} seconds old and so is not a temporary law.

11. Beebee 2000: 547.

12. Although, Armstrong says, a universal cannot exist uninstantiated.

13. Armstrong 1983: 79–80, 100.

14. Armstrong 1997: 257–58.

15. Scientific essentialism (Ellis 2001) holds that the laws are metaphysically necessary; the laws in which a causal power or natural kind figures must be laws in any world in which that power or kind exists. Moreover, a world's essence fixes what kinds and powers exist there (Ellis 2001: 275–76). Therefore, a world's laws appear unchangeable on this account. (Perhaps an essentialist might leave room for changing laws by allowing a world's essence to specify certain kinds as natural before a given moment and other kinds as natural thereafter. However, some arguments for essentialism presuppose that laws must be immutable; see Shoemaker [1998].)

16. In relativistic physics, velocity-boost symmetry involves invariance under Lorentz transformations, whereas in classical physics, it involves invariance under Galilean transformations.

17. If we had some independent way to distinguish fundamental from derivative laws, then we could perhaps express the principle as "All *fundamental* laws are time-displacement symmetric." (Alternatively, perhaps fundamental laws are distinguished from derivative laws partly by their exhibiting various symmetries and other features figuring in "meta-laws.")

I have throughout underdescribed the transformations invoked by the various symmetry principles under discussion. Consider "All emeralds are grue" (where "grue" is defined as "green before the year 3000, blue otherwise"). No time appears *explicitly* in "All emeralds are grue." Accordingly, time-displacement symmetry might be understood as involving invariance under a transformation that takes "All emeralds are grue" into itself. On the other hand, time-displacement symmetry might be construed instead as involving invariance under a transformation that takes "All emeralds are grue" into "All emeralds are green before the year 3000 + a, blue otherwise." These constitute different symmetry principles. To specify a given transformation, it is not enough to say "$t \rightarrow t + a$." We must first specify the form in which the law is to be expressed (including the predicates that are to be used) before the replacement of "t" by "$t + a$."

Here is another example (from John Roberts): suppose it is a law that there is a uniform "gravitational" field through all space, everywhere pointing in the x direction. This law would seem to be invariant under arbitrary spatial displacement since the field is not a function of absolute place. However, instead of invoking a field, the law might be expressed in terms of a potential (see note 26) decreasing uniformly with x. Since x figures explicitly in the potential, the replacement of "x" by "$x + a$" in this version of the law does not leave it unchanged. In fact, it is the potential's (strictly speaking, the Lagrangian's) rather than the field's invariance under various transformations that is linked (in the Hamiltonian framework) to various conservation laws. Again, to specify a particular spatial-displacement symmetry principle, one must specify the particular transformation at issue—or, equivalently, the particular form in which the laws (collectively) must be expressed (for instance, in terms of fields or potentials) before undergoing replacement of "x" by "$x + a$."

I just mentioned the famous correspondence between space-time symmetries and conservation laws. Given a certain fundamental dynamical law (Hamilton's principle; see note 26) and some other conditions (such as that all forces are associated with potentials; again see note 26), energy is conserved if the system's Lagrangian (specifically, its Lagrangian equal to its kinetic energy T minus its potential energy U—see below) exhibits time-displacement symmetry. When I refer to the *laws* being time-displacement symmetric, I must (as we have just seen) be specific about the particular transformation under which the laws are invariant. Regarding first-order laws consisting of Hamilton's principle along with various force laws and so forth, where every force is associated with a potential, I mean that once the laws are expressed in terms of what they say the Lagrangian $T - U$ would be for any possible system, then the laws so expressed are invariant under the replacement of t by $t + a$. Thus, these *laws* are unchanged by this transformation exactly when every *Lagrangian* they allow exhibits time-displacement symmetry.

Here I mean every "Lagrangian" that is equal to $T - U$. Although that is the Lagrangian's canonical form, a Lagrangian (in the broader sense of a function that, given to the Euler-Lagrange equations, yields the system's behavior) may take other forms. For instance, systems where forces are not all associated with potentials may be handled within a Lagrangian framework, but by a more general Lagrangian that does not involve the system's potential energy. That such a system has a Lagrangian exhibiting time-displacement symmetry does not entail that such a system conserves energy. Rather, *if* all forces are associated with potentials and we are taking $(T - U)$ as the Lagrangian (and certain other conditions hold, such as Hamilton's principle), then energy is conserved if the system's Lagrangian exhibits time-displacement symmetry. Furthermore, a system for which $T - U$ is well defined (and so can serve as its Lagrangian) can also have its behavior entailed by Hamilton's principle when functions other than $T - U$ serve as its Lagrangian. Such a function may not be time-displacement symmetric even if $T - U$ is time-displacement symmetric and energy is conserved. So it is not the case that given Hamilton's principle and certain other conditions, energy conservation fails if *a* Lagrangian is not time-symmetric (see note 42).

To appeal to a feature of space (its "homogeneity") as explaining why the laws are invariant under arbitrary spatial displacement seems less plausible when what needs to be explained is why the laws are invariant under a *particular* spatial-displacement transformation (e.g., under a transformation of potentials rather than fields).

Although in the main text I do not pause to fully specify the transformations involved in the various symmetry principles under discussion (say, to spell out how a dynamical law not expressible in terms of a potential is transformed by the transformation figuring in space-displacement symmetry), I presume it understood roughly how these transformations would go (for instance, that a law that each body always moves at 5 meters per second in the $+x$ direction is not invariant under arbitrary rotation).

18. Presuming that some laws are metaphysically contingent, contrary to essentialism.

19. A similar distinction could be drawn in the context of civil and criminal law. That there is no U.S. law respecting the establishment of religion is a matter of meta-law (the First Amendment to the U.S. Constitution), not merely a consequence of what U.S. laws there are. Some parts of the Constitution are meta-laws, such as the laws regulating how a bill becomes a law. Other parts are first-level laws, such as Article III, section 3's specification of what qualifies as treason.

20. Yang 1964: 394.

21. Weinberg 1992: 158.

22. Wigner 1985: 700.

23. Van Fraassen says, "Symmetries of the model…are 'deeper' because they tell us something beforehand about what the laws of coexistence and succession can look like" (1989: 223; cf. 188). But equally, the laws of coexistence

and succession tell us something about the symmetries of the family of models. Suppose the laws of coexistence include $F = ma$ and that the electric force exerted by one charged body on another is proportional to r^{-2} before time T and to r^{-3} subsequently, and there are no laws prohibiting the existence of charged particles. These laws tell us that a model in the family, when transformed by $t \to t + a$ for arbitrary temporal interval a, may not yield another model in the family—that the family of models does not exhibit time-displacement symmetry. In describing a family of models, a symmetry principle might for certain purposes be more informative than some law of coexistence or succession—but for other purposes, one of those laws might be more informative.

Van Fraassen mischaracterizes an ontological priority as merely epistemic. That symmetries come "before" first-order laws, specifying something about what those laws *can* be like, sounds more appropriate (and accounts for the symmetries' greater "depth") if symmetries are not merely ways of describing the first-order laws, but modally more exalted *constraints* governing those laws. So symmetries are ordinarily characterized in physics. Nobel physics laureate David Gross (1996: 14256) is onto the contrast between byproducts and meta-laws—an ontological contrast—when he contrasts the view of symmetries and conservation laws as mere "consequences of the dynamical laws of nature" with the view that "put[s] symmetry first [by] regard[ing] the symmetry principle as the primary feature of nature that constrains the allowable dynamical laws." Gross takes this view to have been Einstein's "great advance in 1905." But it long predates Einstein (see next note). It is now quite standard (e.g., Baumann 2000: 1–2).

On Van Fraassen's view, once the family of models has been specified, symmetry principles add nothing. But I shall argue that symmetry meta-laws add that certain counterfactual conditionals obtain, just as p's lawhood over and above p's truth adds that p would still have held under various counterfactual circumstances.

24. Hamilton 1834/1940: 112; Lagrange 1811/1997: 180, 190, 212, 233.

25. See notes 17 and 26. For explicit remarks about what's explanatorily prior to what, see Landau and Lifshitz (1976: 13); Gross (1996: 14257); Feinberg and Goldhaber (1963: 45); Wigner (1954b: 199). Nearly any physics textbook contains similar remarks.

What about mass conservation? ("The zeroth law of motion, so basic to classical mechanics that Newton did not spell it out explicitly, is that mass is conserved" [Wilczek 2004: 11]). It follows from momentum conservation holding in all inertial frames. Take momentum conservation holding in one inertial frame: for some constant quantity C, $\sum m_i \mathbf{v}_i = C$. (A more rigorous argument would have to include the momentum carried by various fields.)

Consider another inertial frame, moving with velocity V relative to the first inertial frame: $m_i \to m_i' = m_i$, $v_i \to v_i' = v_i - V$. Then the system's total momentum in the new frame is another constant $C' = \sum m_i' v_i' = \sum m_i (v_i - V) = \sum m_i v_i - V \sum m_i = C - V \sum m_i$. Hence, $\sum m_i$ is constant.

26. The explanation of a conservation law by a symmetry principle presupposes that for any type of force, there is a scalar function of position (a "potential" V), determined by the bodies' position coordinates and time, such that the force F of this type on a body with unit "charge" (of the kind relevant to this type of force) at a given location is in the direction from that location in which the potential diminishes most sharply and has a magnitude equal to the potential's slope in that direction. (In other words, $F = -\text{grad } V$.) That is, the explanation presupposes that the "work" needed to put the system into a given configuration (from an arbitrarily selected starting configuration) is independent of the path through state space to that configuration (from the starting configuration). In other words, the system must have a well-defined potential energy. (The body's contribution to the system's potential energy is V times the body's charge.) Under these conditions, an isolated system (which could be the entire universe) of point bodies 1 through N (the momentary state of which consists of the bodies' coordinates $q_1, q_2, \ldots, q_{3N}$ and their instantaneous rates of change $q_i' = dq_i/dt$) has a well-defined Lagrangian $L(q_1, q_2, \ldots, q_{3N}, q_1', \ldots, q_{3N}', t)$ equal to the system's kinetic energy minus its potential energy (see note 17).

Hamilton's principle (presumed to be the fundamental dynamical law governing the system's behavior) says that the system's actual path through state space from its state at time t_1 to its state at time t_2, as compared to nearby alternative paths between those same states at t_1 and t_2, is such that the system's $S = \int L\, dt$ is "stationary" (i.e., is a maximum, minimum, or saddle point). That is, S's variation over small variations in the path vanishes for the actual path. By the calculus of variations, S is stationary only if for each coordinate q_i, the system's path satisfies the Euler-Lagrange equation: $d/dt\, (\partial L/\partial q_i') - \partial L/\partial q_i = 0$. These 3N equations for 3N unknown functions $q_i(t)$ suffice to determine the system's path given L and the initial q_i and q_i'.

Within this framework, the familiar conservation laws are derivable from symmetry principles. The conservation laws are not entailed by the symmetry principles alone; the explanation presupposes in addition not only that Hamilton's principle is (or the Euler-Lagrange equations are) the fundamental dynamical law, but also various other conditions, such as that every force is associated with a potential that is independent of velocity. (I lump together these conditions as "the Hamiltonian framework.")

However, these derivations can all be run in reverse. (I think that the reason why one direction is typically presented and the other omitted from physics textbooks is that symmetry principles are thought to explain conservation laws, not the reverse.) Here is a simple example. Consider a system of two point bodies (masses m_1 and m_2). Let's derive the symmetry of the system's equation of motion under small spatial displacement in the x direction from the conservation of the x-component of the system's linear momentum. Where v_{1x} is the x-component of body #1's velocity and F_{1x} is the x-component of the force on body #1, Newton's second law of motion yields

$$d/dt \, (m_1 v_{1x}) = F_{1x}.$$

As presupposed earlier, F_{1x} can be expressed in terms of the system's potential energy $U(r_1, r_2, t)$:

$$F_{1x} = -\partial U(r_1, r_2, t)/\partial x_1.$$

Then body #1's equation of motion is

$$d/dt \, (m_1 v_{1x}) = -\partial/\partial x_1 \, [U(r_1, r_2, t)].$$

Suppose we displace the system by a small distance a along the x axis:

$$x_1 \to X_1 = x_1 + a$$
$$x_2 \to X_2 = x_2 + a$$
$$t \to T = t$$
$$m_1 \to M_1 = m_1$$
$$v_{1x} \to V_{1x} = v_{1x}.$$

Then body #1's equation of motion is invariant under this transformation if

$$d/dT \, (M_1 V_{1x}) = -\partial/\partial X_1 \, [U(r_1 + a, r_2 + a, T)].$$

By the transformations, this holds exactly when

$$d/dt \, (m_1 v_{1x}) = -\partial/\partial x_1 \, [U(r_1 + a, r_2 + a, t)].$$

For small a, we can use the Taylor expansion

$$U(r_1 + a, r_2 + a, t) = U(r_1, r_2, t) + a[\partial U(r_1, r_2, t)/\partial x_1 + \partial U(r_1, r_2, t)/\partial x_2].$$

So body #1's equation of motion is invariant under this transformation if

$$d/dt(m_1 v_{1x}) = -\partial/\partial x_1 \, [U(r_1, r_2, t)] - a \, \partial/\partial x_1 \, [\partial U(r_1, r_2, t)/\partial x_1 + \partial U(r_1, r_2, t)/\partial x_2].$$

This holds (considering the original equation of motion) if

$$\partial U(r_1, r_2, t)/\partial x_1 + \partial U(r_1, r_2, t)/\partial x_2 = 0,$$

that is, if

$$d/dt \, (m_1 v_{1x}) + d/dt \, (m_2 v_{2x}) = 0.$$

Hence, body #1's equation of motion is symmetric under small spatial displacement in the x direction if

$$d/dt \, (m_1 v_{1x} + m_2 v_{2x}) = 0,$$

that is, if the x-component of the system's total linear momentum is conserved.

Conversely, if U is spatial-displacement invariant, then $U(r_1, r_2, t) = U(r_1 + a, r_2 + a, t) = U(r_1, r_2, t) + a[\partial U(r_1, r_2, t)/\partial x_1 + \partial U(r_1, r_2, t)/\partial x_2]$, so $0 = \partial U(r_1, r_2, t)/\partial x_1 + \partial U(r_1, r_2, t)/\partial x_2 = d/dt(m_1 v_{1x} + m_2 v_{2x})$. Within the Hamiltonian framework, then, the space-time symmetry principle entails the associated conservation law, but the conservation law also entails the symmetry principle.

27. Einstein 1961: 43, my emphasis.

28. Einstein 1954: 329, my emphasis. Compare Earman (1989: 155): "STR is not a theory in the usual sense but is better regarded as a second-level theory, or a theory of theories that constrains first-level theories."

29. Penrose 1987: 24 (his emphasis), cf. 21.

30. Wigner 1972: 10.

31. "[E]ven though we have no catalog of the possible measurements and of the laws of nature...we have reason to believe that we know the abstract group of invariances. This statement amounts to the claim that we know something about the structure of the laws of nature...even though we do not know the laws of nature themselves" (Feynman 1967: 94, cf. Houtappel, Van Dam, and Wigner 1965: 602). That familiar forces are symmetric under spatial reflections was widely considered good evidence that the weak nuclear force is, too, before any phenomena involving that force had been examined for mirror-reflection symmetry. Accordingly, scientists were surprised when parity violations were found (Wigner 1984: 594; Gardner 1964: 239–42).

32. Wigner 1964: 958.

33. Planck in 1887: "If today a quite new natural phenomenon were to be discovered, one would be able to obtain at once from [energy conservation] a law for this new effect, while otherwise there does not exist any other axiom which could be extended with the same confidence to all processes in nature" (Pais 1986: 107–8). Likewise, Feynman says that we are "confident that, because we have checked the energy conservation here, when we get a new phenomenon we can say it has to satisfy the law of conservation of energy" (1967: 76). The reluctance of physicists early in the twentieth century to regard radioactive emission as violating energy conservation suggests that they did not think that energy conservation holds merely as a consequence of the kinds of forces there happen to be. (For more on lawhood's relation to inductive confirmation, see my 2000: 111–59.)

Newton's third law (that any two bodies exert equal and opposite forces upon each other), that any force on a body is exerted by another body, and Newton's second law together logically entail momentum conservation. If (in Newton's physics) this derivation is explanatory and momentum conservation joins Newton's second law in a proper subset of Λ possessing subnomic stability, then this set also contains Newton's third law and that all forces on bodies are exerted by other bodies. (But in classical physics, Newton's third "law" is actually violated by electromagnetic interactions, which are retarded. The momentum conservation law that follows from symmetries includes terms for the momentum in the electromagnetic field and so holds despite Newton's third law being violated. See my 2002: 114–15.) Analogous remarks apply to energy conservation, angular momentum conservation, and (playing the role of Newton's third law and that every force on a body is exerted by another body) that all forces are central forces. "When most textbooks come to discuss angular momentum, they introduce a fourth law [of motion], that forces between bodies are directed along the line that connects them. It is introduced in order to 'prove' the conservation of angular momentum" (Wilczek 2004: 11). (This "law" is violated in classical physics by magnetic forces.)

34. Feynman 1967: 59, 83.

35. Wigner 1972: 13. Likewise Bergmann remarks (1962: 144) that the conservation laws are "general laws applying uniformly to every assembly of mass points regardless of the particulars of the force laws."

36. "[I]nvariance principles can be formulated only if one admits the existence of two types of information [:] initial conditions and laws of nature. It would be very difficult to find a meaning for invariance principles if the two categories of our knowledge of the physical world could no longer be sharply separated" (Houtappel, Van Dam, and Wigner 1965: 596).

37. I argued that a nomically stable set's members possess a species of natural necessity. Some of its members are sub-nomic. Do they then qualify as necessary despite failing to figure in a nonmaximal sub-nomically stable set—contrary to my claim in chapter 2 that for each variety of genuine necessity, the sub-nomic truths possessing it form a sub-nomically stable set? No: I have just shown that for any nomically stable set, its sub-nomic members form a sub-nomically stable set.

38. In the course of refining NP in chapter 1, we considered what would have been the case, had energy conservation not been a law. We noticed that this supposition is logically consistent with all of the sub-nomic m's (taken together) where it is a law that m, but it is not the case that all of those m's would still have held under that supposition. Accordingly, we refined NP to demand that all of those m's would still have held under any *sub-nomic*

supposition that is logically consistent with all of those m's (taken together). Since the supposition that energy conservation is not a law is *not* sub-nomic, energy conservation's failure to be preserved under that supposition does not violate NP. However, Λ^+'s nomic stability suggests a different response we could have given to the problem: instead of requiring that the supposition be sub-nomic, we could have required that it be nomic or sub-nomic but logically consistent with all of the truths of the form "It is a law that m" and "It is not a law that n" (taken together). This requirement would again have rendered irrelevant any behavior under the supposition that energy conservation is not a law.

39. Wigner 1954a: 437–38.

40. Consider the sub-nomically stable proper subset ϕ of Λ containing the conservation laws and fundamental dynamical law, and so forth—but not the force laws. It might appear that "It is a law that energy is conserved," "It is a law that momentum is conserved," and so forth, along with the lawhood of the fundamental dynamical law form a nomically stable set ϕ^+ (that includes their logical consequences among the nomic and sub-nomic claims, such as various symmetry principles). Indeed, ϕ^+'s nomic stability might seem required to capture the truth of such counterfactuals as "Had Coulomb's law not been a law of nature, then momentum conservation would still have been a law," since this counterfactual's truth is not required by ϕ's sub-nomic stability; neither its antecedent nor its consequent is a sub-nomic claim. (Thanks to John Roberts for suggesting this thought.)

However, ϕ^+ lacks nomic stability. For example, the supposition of the force majeure law is logically consistent with the conservation laws together with $F = ma$. (Once again, they all hold—some vacuously—in a universe where it is a law that there is always just a single particle with constant mass moving uniformly forever.) But ϕ^+ is not preserved under this supposition. As another example, consider the supposition that it is a law that the sum of each body's $(mv)^{1/2}$ is a conserved quantity. This supposition is logically consistent with ϕ^+. (In a universe where it is a law that there is nothing but a single point body of constant mass moving uniformly forever, it is a law that $\Sigma(m_i v_i)^{1/2}$ is conserved, and all of the actual conservation and dynamical laws are still laws, too.) But (in at least some conversational contexts, it is true that) had this supposition held, some actual conservation laws would not still have held; momentum or energy conservation would presumably have been replaced by the law of $\Sigma(m_i v_i)^{1/2}$ conservation. Hence, ϕ^+ lacks nomic stability.

Furthermore, counterfactuals like "Had Coulomb's law not been a law of nature, then momentum conservation would still have been a law" can be accounted for solely by ϕ's sub-nomic stability. Recall that a set's sub-nomic stability requires the truth of various nested counterfactuals. If the strata of natural laws among the sub-nomic truths are cashed out as the nonmaximal sub-nomi-cally stable sets, then thanks to those nested counterfactuals (as I explained in chapter 1), ϕ's sub-nomic stability demands that had Coulomb's law been false, then momentum conservation would still have been *a law*. Furthermore, had Coulomb's law been false, then Coulomb's law would not have been a law (since laws, like accidents, are truths). In addition, just as there would have been a body accelerated from rest to beyond the speed of light had there been no law prohib-iting such a thing, so likewise had Coulomb's law not been a law, it would not have been true. (At least that's true in typical contexts.) Thus we have (in those contexts) $p \; \square\!\!\rightarrow q$ (Coulomb's not a law $\square\!\!\rightarrow$ Coulomb's false), $q \; \square\!\!\rightarrow r$ (Cou-lomb's false $\square\!\!\rightarrow$ momentum conservation still law), and $q \; \square\!\!\rightarrow p$ (Coulomb's false $\square\!\!\rightarrow$ Coulomb's not a law), from which it logically follows (by a principle of counterfactual logic; see Lewis [1973: 33]) that $p \; \square\!\!\rightarrow r$ (Coulomb's not a law $\square\!\!\rightarrow$ momentum conservation still law), which was our target.

41. If it is a law that there are no bodies, then vacuously each body moves always in the $+x$ direction, but rotational symmetry is not violated since the $+x$ direction is not thereby privileged. (Recall from the previous section that a symmetry principle pertains to the laws as a whole.)

Objection to Λ^{meta}'s nomic stability: It requires that "Had it been a law that each body moves always in the $+x$ direction, then there would have been no bodies" is true (in every context). Is this a good reason to believe that the sym-metry principles are not meta-laws (or to reject my account of what it would take for them to be meta-laws)? I don't think so. Admittedly, even if we believe that symmetry principles are meta-laws, it might initially sound more plausible to say, "Had it been a law that each body moves always in the $+x$ direction, then there would have been less symmetry in the laws," rather than "...then there would have been no bodies." However, this impression may derive partly from our failure to bear in mind, while we are evaluating these counterfactuals, that symmetry meta-laws *constrain* the first-order laws, or partly from our mis-takenly thinking that this counterfactual's antecedent includes that there are bodies and so is logically inconsistent with the symmetry principles (just as we might find ourselves tending to deny that it is a law that each ghost moves always in the $+x$ direction, mistakenly failing to consider that this law follows from the law that there are no ghosts). Admittedly, in explaining space-time symmetries to a class, I might say, "Had it been a law that ordinary matter always seeks its natural place at the center of the universe, then the laws would not have respected space-displacement symmetry" (rather than "...then there would have been no ordinary matter"), which runs contrary to Λ^{meta}'s nomic stability. But the example that I am giving the class might be properly expressed not by a counterfactual at all, but rather by something indicative, such as "Under Aristotelian laws, the symmetries are more meager."

Lewis's Best System Account could nicely characterize meta-laws as belonging to the Best System of truths about the Best System of truths about the Humean mosaic. Of course, since the Best System Account does not link lawhood to stability, it does not entail that if Λ^{meta} contains the meta-laws, then had it been a law that ordinary matter tends to go to the universe's center, space-displacement symmetry would still have held and so there would have been no ordinary matter. This result might seem attractive. However, I regard my account as better capturing how meta-laws *constrain* the first-order laws (and how first-order laws constrain the sub-nomic facts).

42. As I mentioned in note 17, a system for which the canonical Lagrangian $T - U$ is well defined can also have its behavior entailed by Hamilton's principle when certain functions other than $T - U$ serve as its Lagrangian. Such a Lagrangian may not be time-symmetric, but if the system conserves energy, then that Lagrangian exhibits some symmetry that (under the relevant conditions) entails energy conservation. Why, then, is energy conservation in this case explained by the standard Lagrangian's time-displacement symmetry rather than by the nonstandard symmetry of one of these nonstandard Lagrangians? The reason, I suggest, is that a given nonstandard Lagrangian's symmetry is an isolated accident, whereas the standard Lagrangian's symmetry is required as a matter of meta-law. That is: One explanation of the conservation law (and hence of energy's conservation in this particular case) proceeds from a symmetry meta-law. Suppose that the first-order laws consist of Hamilton's principle along with various force laws and so forth, every force associated with a potential. There is no meta-law requiring that these laws be such that every system they allow have a certain kind of nonstandard Lagrangian: one exhibiting the same symmetry exhibited by the given nonstandard Lagrangian. In contrast, there is a meta-law requiring that these laws be such that every system they allow have a Lagrangian $T - U$ that is time-displacement symmetric.

43. Other varieties of meta-laws are discussed in my 2000.

44. Lewis (1999: 229) uses this isotope to pose difficulties for frequentist accounts of chance.

45. Lewis 1986: 128, cf. 125; 1999: 233–34.

46. A system can include (50%) and (1) without violating the "requirement of coherence"—by also including "No ^{346}Un atom ever exists."

47. Instead of a system containing (50%), I might just as well have considered a system containing some other statistical generalization that, when combined with the universe's history of elite-property instantiations, entails (50%).

48. Because E is about to be conditionalized upon, and if E is logically impossible, then (if the agent is logically omniscient) $cr(E) = 0$, and so $cr(A|E)$ is undefined according to the standard definition of $cr(A|E)$ as $cr(A\&E)/cr(E)$.

49. For E to logically entail A, E did not need to include (1)'s lawhood, merely (1)'s truth. But as we will shortly see, (1) gives the appearance of qualifying as admissible only by entering E as a deterministic law.

50. The literature contains occasional remarks such as "given that information about the laws is admissible" (Schaffer 2003: 31).

51. In effect, this argument instantiates Lewis's argument that by PP, chances must obey the probability calculus.

52. Lewis 1999: 234.

53. Lewis (1986: 120) says that in a deterministic universe, all chances are 0% and 100%. Lewis thereby appears to assume that if (1) is a law, then (100%) is true.

54. Indeed, perhaps (1)'s lawhood even entails that (100%) is *false.* That (1) is a law logically entails that it is (naturally) impossible for a ^{346}Un atom to live beyond 7 μs after its creation. But perhaps (100%), by entailing that a ^{346}Un atom's decay within 7 μs is the outcome of a chancy process, logically entails that a ^{346}Un atom's survival beyond 7 μs *is possible,* though it has no finite likelihood—just as a fair coin (or even a coin with a head on one side but biased 100% in favor of tails) can possibly land heads on each toss in an infinite sequence, though it has no finite nonzero likelihood of doing so. As (1)'s lawhood logically precludes (50%)'s truth, so perhaps it also logically precludes (100%)'s truth.

Lewis avoids this result by embracing infinitesimal chances (see Lewis 1986: 88–90, 125, 175–76). On his view, it is possible for a ^{346}Un atom to live beyond 7 μs as long as there is even an infinitesimal chance of its doing so, whereas (100%)—and not an infinitesimal less!—entails that such a thing is impossible. Yet we might feel reluctant to exploit nonstandard probabilities to capture the relation between chancy facts and deterministic laws.

55. It is an instance of the objective-chance counterpart of the (Special) Reflection Principle for credence (Van Fraassen 1984).

56. Lewis (1999: 230) regards the question of whether there could be "lawless chances" as "spoils to the victor."

57. For more on irreducibly second-order chances (i.e., chances of chances), see my 2006.

58. Here I appeal to the familiar rationale for denying that the principle of conditional excluded middle applies to counterfactuals—see Lewis 1986: 329–31.

59. See chapter 1, note 24 on "might" counterfactuals; here I am assuming that at least in some contexts, the "might = not-would-not" relation holds.

60. Lewis 1986b: 126.

61. An objection to my *reductio:* Suppose not only that (1) is a law and (50%)'s nonvacuous truth is logically consistent with all of the *m*'s where it is a

law that m, but also that (50%) is nonvacuously true. (It would beg the question to object that (50%) cannot be non-vacuously true if (1) is a law. That's the relation we are trying to explain!) Then were (50%) nonvacuously true, (1) would still be true (since both *are* true!). But that is precisely the conditional I held to be false in my *reductio*.

One reply: Consider instead the conditional "Had there been 600 [346]Un atoms [more than there actually are, let's suppose] while (50%) remained the case, then (1) would still have been true." If its antecedent is logically consistent with all of the m's where it is a law that m, then (considering (1)'s lawhood) this conditional must be true—but (for the same reason as before) it is false (in some contexts, at least). Hence (if no law constrains the number of [346]Un atoms), (50%)'s nonvacuous truth must be logically inconsistent with some m where it is a law that m.

A more radical reply: If at t exactly one [346]Un atom is created and it has a 50% chance of decaying within $7\,\mu s$, then even if it decays within $7\,\mu s$, there are certain contexts where "Were exactly one [346]Un atom created at t, then it would decay within $7\,\mu s$" is false. The atom might decay, but it might just as well not do so; later chance outcomes have no influence (in these contexts) over the closeness of possible worlds. The actual world is only tied (in these contexts) for the closest possible world where exactly one [346]Un atom is created at t. In other words, I reject "Centering" when there are objective chances. See also note 64 and chapter 4, note 1.

62. Related arguments suggest that the laws must be "complete"—see chapter 4, section 4.8.

63. Current elementary particle physics may turn out not to be true. Suppose it is true (at least in this respect), for argument's sake.

64. At least, I claim, there are some contexts like this—where the actual future outcomes of chance processes have no influence on the closeness of various possible worlds to the actual world. There may be other contexts where "Were I to flip the coin at noon today, it would land 'heads'" is true (unbeknownst to me) before noon if I actually flip the coin at noon and it lands "heads."

Chapter 4

1. In my view, p does not even help to make it the case that $q \; \square\!\!\rightarrow p$ when q obtains. Indeed, I think there are contexts where both q and p obtain, but $q \; \square\!\!\rightarrow p$ does not—contrary to "Centering" in the Stalnaker-Lewis possible-worlds account of counterfactuals. (I foreshadowed this point earlier—e.g., in chapter 3, notes 61 and 64.)

For instance, suppose that an X-on (a certain species of elementary particle) has a half-life of 7 seconds. (I assume that its decay is an irreducibly statistical process; the half-life does not reflect our ignorance of some "hidden
variable" that determines when it will decay.) Suppose that atoms of a certain
element Q are radioactive, with a half-life of 100 years, and when a Q atom
decays, it must emit a single X-on. Suppose we have before us a Q atom. Consider "Were this Q atom to decay sometime in the next 100 years, then the
X-on it produces would decay within 7 seconds thereafter." In some contexts,
this conditional is false; if the Q atom were to decay sometime in the next 100
years, then the resulting X-on might decay within 7 seconds of its creation (it
would have a 50% chance of doing so), but it might just as well fail to do so.
This is true (in those contexts) even if the Q atom actually decays in the next
100 years and the daughter X-on actually decays within 7 seconds of being
created. Here, then, we have an example where q and p obtain but $q \; \square\!\!\rightarrow p$ does
not. (Cf. Bennett 2003: 234, 240–41.)

2. On the other hand, some Schoolmen (such as Francisco Suárez) thought
that there are primitive subjunctive facts regarding the actions that would be
taken freely by certain agents. The agent's desires, intentions, and character
together with the natural laws cannot be responsible for those facts on pain of
undermining the agent's freedom. Some philosophers, then, apparently believe
in some primitive subjunctive facts.

3. Our intuitions about the laws are sometimes much more straightforward than our intuitions about which counterfactuals are true and which are
false—especially when it comes to nested counterfactuals. But it does not follow that a reduction of lawhood to subjunctive facts lacks intuitive appeal.
(Furthermore, many counterfactual conditionals were known to be true long
before any relevant laws had been discovered.)

4. Goodman's classic discussion, a model of philosophical craftsmanship, is
chapter 1 of his *Fact, Fiction, and Forecast* (Goodman 1983).

5. We might instead try locating singular causal relations at the bottom.
Nevertheless, Goodman showed that it is not obvious that subjunctive
conditionals are made true exclusively by laws and nonsubjunctive facts.
Though we may be tempted to think laws partly responsible for the truth of
"struck $\square\!\!\rightarrow$ lit," are we equally tempted to think laws partly responsible for
the truth of "struck $\square\!\!\rightarrow$ dry"? It may also be tempting to think that "Had p
obtained, then q would have obtained" means "...then q would *have to* have
obtained," where "have to" invokes natural necessity. But as we saw in chapter
1, note 24, "would have" is not synonymous with "would have to have."

Admittedly, a great deal more than I manage here needs to be said about
"subjunctive facts." How do they differ from other kinds of facts? Is there anything especially "subjunctive" about them (or only about what they make

true)? Why do they relate to one another so as to obey the logic of subjunctive conditionals? Is there a distinct primitive subjunctive fact for every subjunctive conditional and context where that conditional is true? Questions analogous to some of these can also be raised regarding sub-nomic facts—without throwing any suspicion on their ontological bona fides. Fortunately, no feature of my account of laws turns on giving certain answers rather than others to these questions.

6. Putnam 1990: 87–88.

7. In contrast, Van Fraassen (1989) holds that lawhood does not figure in a rational reconstruction of science, and he argues against the objectivity of lawhood partly by appealing to the fact that counterfactual conditionals lack objective truth-values. See note 9.

8. Putnam (1990: 87–88) draws this analogy.

9. From the context sensitivity of such counterfactuals, Van Fraassen argues that "science by itself does not imply" them (1989: 35) since "scientific propositions are not context-dependent in any essential way" (1980: 118): "Because science cannot dictate what speakers decide to 'keep constant' it contains no counterfactuals" (1977: 149). But doesn't science tell us *in a given context* whether or not some counterfactual conditional is true? Some scientific claims plainly express different propositions in different contexts. How close Jones's height must be to exactly six feet, for "Jones is six feet tall" to be true, differs in different contexts—without preventing the truth of the claim about Jones's height from being ascertained scientifically in a given context.

10. Armstrong 1997.

11. Bigelow 1988: 130–33.

12. Lewis 1999b: 218.

13. Advocated in Maudlin 2007.

14. Of course, it might be suggested that counterlegals and counterlogicals should be treated as special cases rather than by the very same account that deals with "ordinary counterfactuals." This seems to me an unfortunate bullet to have to bite.

15. Earman (1986: 100) gives a different example. By the way, I use a capital "F" because I do not want to assume in the upcoming discussion that the fact responsible for p's necessity is sub-nomic. Moreover, I take F, p, C and so forth in the upcoming discussion to be hypothetical states of affairs that obtain (or not) in a given possible world purely in virtue of how things are *in that world*. So, for example, "$C \,\square\!\!\rightarrow p$" cannot be "Had I been 6 feet 7 inches tall, then I would have been one foot taller than I actually am," since my being (in a given possible world) one foot taller than I actually am is a matter not just of how things are in that possible world (my being 6'7" there), but also of how things are in the actual world (my being 5'7").

16. Nordenskiold 1936: 398. In 1845, T. H. Huxley found this passage pithy enough to place atop his student notepad (Desmond 1997: 28).

17. In chapter 1, I suggested that a natural law is like a mathematical truth in that it possesses explanatory power by virtue of its necessity. For example, the fact that 23 cannot be divided evenly by 3 explains why it is that every time mother tries to divide 23 strawberries equally among her three children without cutting any (strawberries), she fails.

18. Blackburn 1993: 53.

19. Van Fraassen 1980: 213. Van Fraassen's view seems to be that this problem is unsolvable. He writes: "What exactly is this criterion, that laws must explain the phenomena? . . . What makes laws so well suited to secure us this good? When laws give us 'satisfying' explanations, in what does this warm feeling of satisfaction consist? There are indeed philosophical accounts of explanation, and some mention laws very prominently; but they disagree with one another, and in any case I have not found that they go very far toward answering *these* questions" (1989: 31). Van Fraassen believes that lawhood is not a useful concept for a rational reconstruction of science and that offering "scientific explanations" is not part of doing science (but merely an extracurricular activity in which some scientists are engaged). I side with Steven Weinberg: "To tell a physicist that the laws of nature are not explanations of natural phenomena is like telling a Tiger stalking prey that all flesh is grass" (1992: 28–29).

20. Perhaps none of its lawmakers individually *had to be,* but at least there had to be lawmakers sufficient to make it a law.

21. Interestingly, Henle seems sensitive to this puzzle. Immediately after identifying laws as explaining phenomena by making them necessary, he admits, "It is true, even these laws offer no explanation as to the ultimate grounds" (Nordenskiold 1936: 398). But then how *do* they manage to explain? If the laws merely make various facts necessary *given* the laws, then why is that enough to explain those facts? (Recall the *Star Trek* example from chapter 2.)

22. The puzzle is *not* motivated by the thought that every contingent fact must have an explanation. The puzzle concerns the source of a law's necessity (and, hence, of its explanatory power). It does not presuppose that there are no contingent unexplained explainers.

23. It seems arbitrary to exclude such subjunctive facts from sub-nomically stable sets if the sub-nomic facts include various dispositional facts. See two paragraphs below and chapter 1, notes 23 and 25.

24. Goodman 1983: 40.

25. Bennett (2003: 167–68) offers this example: "Jones was not careless when he threw the lighted match onto the leaves. He knew that the leaves were too damp to ignite. If it had been the case that if he were to throw the match onto the leaves a forest fire would ensue, then he would have known this

was the case and not thrown the match onto the leaves." Later I will argue that a fact about a body's instantaneous velocity at t is a fact about what the body's trajectory would be like, were the body to remain in existence after t. Hence, "Had the marble's speed at t been 10 centimeters per second" conceals a subjunctive conditional.

Consider a match that is wet (so had it been struck, it would not have lit) but otherwise in propitious conditions. Had it been true that the match would have lit had it been struck, the match would have been dry. Had the match been dry, then the match would have lit, had it been struck, and the actual laws of nature would still have been laws. Therefore, had it been true that the match would have lit had it been struck, then the actual laws of nature would still have been laws. I shall now include such counterfactuals in my account of the laws' relation to counterfactuals.

Of course, the truth-values of counterfactuals with antecedents involving counterfactuals are context-sensitive. In one conversational context, it might be accurate to say, "If you would get a million dollars were you to touch Jason's head, then everyone would be chasing Jason," whereas in another conversational context, it might be accurate to say, "If you would get a million dollars were you to touch Jason's head, then not everyone would be chasing Jason; some people would be touching other people's heads to see if that would work, too."

Question: How is $(p \: \square\!\!\rightarrow q) \: \square\!\!\rightarrow m$ to be understood in a nonbacktracking context if $(p \: \square\!\!\rightarrow q)$ is a backwards-directed counterfactual? What would it be for $(p \: \square\!\!\rightarrow q)$ to be true in a nonbacktracking context? For instance, if $(p \: \square\!\!\rightarrow q)$ is "Had I worn an orange shirt this morning, then Lincoln wouldn't have been assassinated," then $(p \: \square\!\!\rightarrow q)$ is false in a nonbacktracking context, and so how can we—in a nonbacktracking context—entertain the counterfactual supposition that it is true?

A rough answer: In a nonbacktracking context, "Had I worn an orange shirt this morning, then Lincoln wouldn't have been assassinated" is true exactly when Lincoln was not assassinated. So the counterfactual antecedent "Had it been the case that Lincoln wouldn't have been assassinated had I worn an orange shirt this morning...," entertained in a nonbacktracking context, amounts to "Had Lincoln not been assassinated...."

26. See the end of note 23, chapter 1. It is unnecessary to require that a "Stable" set's members all be true: to be Stable, Γ must be logically consistent (since otherwise there is no ϕ where $\Gamma \cup \{\phi\}$ is logically consistent) and so it must be true (in any context) that any member would hold under the supposition of any logical truth p, which precludes falsehoods from membership.

27. This recipe for causing trouble can also be deployed against an account that takes lawhood as primitive. (See section 4.1.)

28. See Ellis 2001: 278 (quoted in chapter 2); 2005: 76.

29. The same may be said even of Lewis's drastically different account of counterfactuals, according to which (if the actual universe is deterministic) had I worn an orange shirt, the laws of nature would have been different. (Lewis, then, denies the claim deemed so innocuous by Ellis: that we should look to a world "belonging to the same natural kind" as the actual world.) Lewis regards the "miracles" (violations of actual law) occurring in some possible world as especially influential in determining that world's "closeness" to the actual world. (See chapter 1, note 29.)

30. Russell 1919: 71.

31. Handfield (2005: 83) and Ellis (2005: 78) make suggestions along these lines.

32. If essentialism is correct, then strictly speaking, there are no electrons in such a world—only particles much like electrons but with greater electric charge. But presumably the essentialist would say that for the sake of simplicity, we continue speaking of "electrons," "protons," and so forth in connection with the "possible world" invoked by this counterfactual conditional.

33. Ellis 2001: 278. Law primitivism (mentioned in section 4.1) also has few resources with which to account for meta-laws and multiple strata of first-order laws.

34. Indeed, one surprising consequence of my argument in section 4.8 below is that there isn't even a "possible world" where all and only the laws of classical physics rule. (See note 72.)

Although I will appeal to absolute space and time for the sake of simplicity, my argument could easily be extended to relativistic physics; the instantaneous velocity in question would then be relative to a given reference frame. Quantum mechanics is quite another matter.

35. Here I assume, for simplicity, that the body is moving in a straight line. I shall be sloppy about vector notation, allowing context to indicate whether I mean speed (a scalar) or velocity (a vector: speed and direction) by "v," and likewise for other symbols.

36. See Russell 1903/1937: 473; Russell 1917: 84; Salmon 1980: 41; and Salmon 1984: 152.

37. Therefore, I shall disregard the argument (Tooley 1988: 247–48) that the reductive analysis of instantaneous velocity is inapplicable to a body at t_0 whose most recent moment of existence before t_0 is at $t_0 - \Delta t$ (for some finite, nonzero Δt).

Here, from Peter Guthrie Tait, is a typical statement of this traditional interpretation of classical physics: there are "limits" on the motions that are "possible in the case of a particle of matter. These limitations are simple, but very important. The path of a material particle must be a *continuous* line. (A gap in it would imply that a particle could be annihilated at one place and reproduced at another.)" (Knott 1911: 233).

38. For some examples of this traditional interpretation of classical physics, see Walton 1735: 47; Emerson 1768: v and xi; Maclaurin 1748/1971: 54 and 113-14; Thomson and Tait 1888: 242 and 385. (See also the previous note.) Philosophers, too, generally presume something like this traditional causal interpretation, as when Lewis takes a car crash as having among its causes the position and velocity of the car a split second before the impact (Lewis 1986b: 216; cf. Hempel 1965: 184, 449).

Of course, philosophers who believe that causal relations are not objective features of reality, or that it is not the business of scientific theories to describe causal relations, might regard it as no defect in the reductive account that it portrays classical instantaneous velocity and acceleration as unable to play their traditional causal roles. Moreover, even if a philosopher believes in the objective reality and scientific relevance of causal relations, she might nevertheless insist that classical mechanics should not be given a causal interpretation or that its correct causal interpretation portrays instantaneous velocity and acceleration as epiphenomenal rather than as playing causal roles. I shall not address these views directly. I shall merely investigate what classical instantaneous velocity and acceleration would have to be like, were they to play the causal and explanatory roles that classical physics is traditionally interpreted as attributing to them (taking them as exemplifying the causal roles of other instantaneous rates of change).

39. I shall assume that the body feels merely external forces shoving it around—that it cannot break apart, ignite, etc.

40. Tooley (1988: 240 and 243), Bigelow and Pargetter (1990: 66), and Arntzenius (2000: 192) briefly sketch arguments in a roughly similar spirit. Walton (1735: 47) and Emerson (1768: v and xi) are careful to distinguish instantaneous velocity from its effect (involving change of place).

41. It might be replied: Yes, instantaneous velocity has no causal role in classical physics. A body's *momentum* does the causal work. By physical law, a body's momentum equals its mass times velocity, but momentum is ontologically distinct from mass and velocity; momentum is ontologically on a par with charge and trajectory. In accordance with Newton's second law of motion (equating the net force on the body at t to the time rate of change at t of the body's momentum), a body's trajectory in the interval $<t_0, t_0 + \Delta t]$ is causally explained by the body's mass, the forces on the body at each moment in the interval $[t_0, t_0 + \Delta t]$, and some initial conditions: the body's position at t_0 and the body's momentum at t_0.

I reply: The causal explanation problem is now reproduced as a puzzle about momentum and its instantaneous rate of change. That rate at t_0 (caused by the net force on the body at t_0) is supposed to be a cause of the body's momentum in $<t_0, t_0 + \Delta t]$. How, then, can momentum's instantaneous rate of change at t_0 be reduced to a certain relation among the body's momenta at instants in t_0's neighborhood?

42. This principle is indifferent to whether e is a fact or an event, and likewise for the other causal relata.

43. The presupposition that a relation's holding is a cause only if the relata are acknowledges relationships as able to be causes. Some philosophers contend that only intrinsic properties (perhaps along with spatiotemporal relations) can be causally relevant. Sometimes this appears as the view that only events can be causes and that events are predominantly intrinsic. (See, e.g., Lewis 1986b: 262.) I want my argument against the reductive account of instantaneous velocity to remain independent of any such controversial premises. However, those who are willing to embrace such premises could put the causal explanation problem in this way: On the reductive view, a body's $v(t_0)$ is not an intrinsic property of the body at t_0. So a body's having a given instantaneous velocity at t_0 is not an event. It therefore cannot be a cause of the body's subsequent trajectory.

44. Tooley (1988: 243) considers and rejects this alternative, though he seems to think that it succeeds at least in avoiding the causal explanation problem.

45. I used the ramp function merely to illustrate the derivative "from below." In classical physics as traditionally interpreted, a body's trajectory cannot be given by $R(t)$ because such a body would have to undergo infinite acceleration at $t=0$. Such a momentarily infinite acceleration cannot be plugged into the causal law governing how a body's final velocity $v(t_2)$ at the end of the interval $[t_1, t_2]$ is caused by its initial velocity $v(t_1)$ and the instantaneous acceleration $a(t)$ that it feels at each moment during that interval: $v(t_2) = v(t_1) + \int a(t)\, dt$ (the integral ranging from t_1 to t_2). If $a(t)$ equals zero at all times except $t=0$, when it becomes infinite, then $a(t)$ cannot be integrated.

Of course, a momentarily infinite acceleration can be a useful approximation. The conservation laws may enable us to solve for the resulting trajectory without having to give a causal explanation of it. There are mathematical devices for representing such an acceleration (the "delta function") and for manipulating it (via "integration in the distributional sense"). But they do not allow a momentarily infinite acceleration to be used within a causal interpretation of classical physics. (See Zemanian 1965: 2.)

The infinite forces required by momentarily infinite accelerations have typically been viewed with extreme suspicion by physicists interpreting classical physics in traditional causal terms. See, for instance, Thomson and Tait (1888: 1) and the end of the passage from Tait that I quoted in note 37: The "limits" on the motions that are "possible in the case of a particle of matter" include that "[t]here can be no instantaneous finite change in the direction, or in the speed, of the motion."

46. Likewise, this principle entails that a body's $v(t_0)$ cannot be a cause of its change of position between t_0 and $(t_0 + \Delta t)$ unless $v(t_0)$ is either a cause of $x(t_0)$ or a cause of $x(t_0 + \Delta t)$.

47. Even if backward causation is admissible in exotic cases, this is not an exotic case. Also E at x_0 and the body's occupying x_0 at t_0, along with the body's mass and charge (and perhaps the absence of other forces), form a *complete* cause of $a(t_0)$. Presumably, then, if $a(t_0)$ is a relation's holding, then *each* relatum must have some member of the complete cause as a cause. But it is difficult to see how this can be.

48. Jackson and Pargetter (1988) and Meyer (2003: 97) have offered proposals along roughly these lines.

49. See Feynman, Leighton, and Sands 1963: vol. 2, p. 21–1.

50. Views along these lines are defended by Tooley (1988) and Bigelow and Pargetter (1990).

51. For example, Arntzenius 2000: 197.

52. Tooley (1988: 238–39) defines "velocity" as whatever property stands in such relations. On Tooley's view, though, it is not essential to whatever property in fact stands in these relations that it does so. That property merely qualifies as velocity in virtue of standing in these relations.

53. I reject "Centering," which would require $p \ \square\!\!\rightarrow q$ if p and q are true. See note 1 above and chapter 3, notes 61 and 64.

54. For example, a world operating according to classical physics (interpreted causally) contains no field that affects trajectory by automatically making any body located within it move at a uniform 2 cm/s, since such a field would occasionally have to produce an instantaneous, finite change in a body's velocity, and as we saw (note 45), this cannot be accommodated within classical physics (interpreted causally).

55. Analyses roughly like mine were offered long ago (though since the late nineteenth century, they have been eclipsed in favor of reductive views). Thomson and Tait (1888: 12) "define the exact velocity [of a point body] as the space which the point would have described in one second, if for one second its velocity remained unchanged." A similar account is offered by Maclaurin, who calls instantaneous velocity a "power" (Maclaurin 1742: 53–55; see Jesseph 1993: 281–82 and Carroll 2002: 66). Walton (1735: 47) calls a body's instantaneous velocity the "Tendency forward in the body." Maclaurin (1748/1971: 104; cf. Hutton 1796: 484) says that "the velocity of motion is always measured by the space that would be described by that motion continued uniformly for a given time." Although these proposals have a subjunctive character, they appear to be circular; in defining a body's instantaneous velocity at t_0, these proposals

appeal to the body's instantaneous velocity's remaining constant over a finite period beginning at t_0. My proposal avoids this problem.

56. Bigelow and Pargetter 1990: 68–70; Tooley 1988: 244.

57. In such a context, it is true that were the body to continue to exist after t_0, its trajectory at t_0 *might* have a well-defined time-derivative from above. However, on my proposal, it does not follow that the body at t_0 might have a $v(t_0)$.

58. Faraday 1858: 560.

59. This notion of completeness is suited only to a universe with absolute time. No matter: I shall suggest later that different law-governed universes have different completeness principles.

60. I shall presume that E is not vague. Otherwise $\mathrm{ch}_T(e)$ might be vague. The demands of "completeness" might be extended to accommodate this possibility.

61. Compare Papineau's (2001: 8) thesis of "the completeness of physics": that all physical occurrences (or, to accommodate quantum mechanics, their chances) are determined by law together with physical prior history.

62. Van Fraassen (1989: 171) credits this aphorism to Oliver Wendell Holmes.

63. Feynman 1967: 151.

64. See chapter 2, note 17.

65. Likewise, the transition rules of a cellular automaton (such as Conway's "game of life") are frequently characterized as its "laws of nature." Such rules are complete: for each of the $2^9 = 512$ possible patterns of occupation of a 3×3 grid, the rules specify whether or not the central square is occupied at the next time step.

Admittedly, if a game's rules are incomplete but no cases ever threaten to fall into a gap, then their incompleteness might never lead to problems. Games where the rules are made up as the players go along may typically have incomplete rules and nevertheless work adequately. Furthermore, even if an ideal game's rules must cover "every possible eventuality," this range may be narrower than the circumstances that are logically consistent with the rules. For instance, the rules of a ball game may be complete yet fail to cover a case where a ball turns into a bird and flies away. (This example was suggested to me by Bill Lycan, who believes it was Wittgenstein's.) That is not one of the "possible eventualities" because the game presupposes certain background conditions that are not entailed by its rules. Presumably, any such "background conditions" for nature's game are among the natural laws.

66. See chapter 2, note 18.

67. Bergmann 1980: 156

68. Earman (1995) contains many passages evincing this attitude (including Bergmann's remark). Concerning singularities in general relativity, however, Earman regards it as "a pious hope that some quantum theory of gravity, yet to be formulated, will contain mechanisms for [their] avoidance" (224).

69. Earman 1995: 94; cf. 65–66.

70. Indeed, even the possibility of a clothed singularity would reveal the laws to be incomplete, since events inside the event horizon would fail to be covered by the laws.

71. Earman 1995: 225.

72. Surprisingly, in a world "governed" by something like the laws of Newtonian physics, initial conditions and laws allow certain events to occur (such as "space invaders" swooping in from infinity and particles spontaneously initiating motion) while leaving room for those events not to occur, and no particular chances are assigned them by the laws and initial conditions (Earman 1986; Hutchison 1993; Laraudogoitia 1996; Norton 2007). But if the laws must be complete, it follows that no such world is possible. Admittedly, a Newtonian world *seems* possible—but perhaps not after we notice that its laws have gaps. (See also the text at note 81.)

73. Lewis's account could be amended to require that any system eligible for the competition for Best be complete—or completeness could join informativeness, simplicity, and fit as desiderata for the "Best." (Loewer [2004: 1118] may be considering the latter.) But it would be better for the laws' completeness to derive nicely from some integral part of the account than to be inserted "by hand."

74. David Armstrong (personal communication) kindly acknowledged that on his account, the laws do not have to be complete—adding, however, that he "should think worse of the world if there actually is 'incompleteness'" (!).

75. For two reasons, each of which is sufficient: (i) Suppose it is a law that m. Since m is a member of $\Lambda°$ and ϕ is logically consistent with $\Lambda°$, m must be preserved under ϕ for $\Lambda°$ to be Stable—so $(\phi \,\square\!\!\rightarrow m)$ must be true. (ii) To be Stable, $\Lambda°$ must consist exclusively of truths (see note 26), and $(\phi \,\square\!\!\rightarrow m)$ is a member of $\Lambda°$ (since ϕ is logically consistent with the cascade of conditionals that had to be true for Λ to qualify as sub-nomically stable)—so $(\phi \,\square\!\!\rightarrow m)$ must be true.

76. Unlike Goodman, I am not presuming that what *makes* $p \,\square\!\!\rightarrow q$ true is that one of these two options holds. I am presuming only that if the counterfactual conditional is true, then one of these two options holds.

77. Let me emphasize why I have considered a conditional with a counterfactual conditional ϕ as its antecedent rather than a more straightforward conditional, such as "Had an A-B interaction occurred in L at 1–2 nm and produced green slime, then. . . ." To suggest that under the latter's antecedent, there would

have been a green-slime law requires presupposing that any event must be covered by a law. To avoid thereby begging the question, I have instead appealed to a conditional having ϕ as its antecedent. I do not presuppose that every counterfactual conditional that is contingently true (having an antecedent logically consistent with the laws) must be covered by a law. Rather (as I mentioned), if $p \,\square\!\!\rightarrow q$ holds, where p is logically consistent with Λ, there may be certain actual accidents f that the context invokes where $(p \,\&\, f)$ suffices to logically entail q. (I do not even presume that a world where nontrivial counterfactual conditionals obtain must have laws.)

78. John Carroll kindly suggested the following reply. Take Tooley's example (1977: 669) of a world with exactly 10 kinds of fundamental particles, and hence 55 kinds of fundamental two-body interactions. Suppose the laws are incomplete: there are laws for 54 of these kinds of interactions (including A-B interactions, at all distances), but none governing the interaction of an X-on with a Y-on. As it happens, no such interaction ever occurs. The 54 laws are utterly dissimilar. Let ϕ be: had an X-Y interaction occurred, then all such interactions would have turned the interacting particles into green slime. Carroll suggests that even if the first step of my argument goes through so that ($\phi \,\square\!\!\rightarrow$ green-slime law), my second step fails to secure that ($\phi \,\&\,$ green-slime law) $\diamondsuit\!\!\rightarrow \sim\!\Lambda$ (holds in some context): considering the tremendous diversity among the 54 laws, no pressure would be put on any of them by the supposition of yet another law unlike each of them (the green-slime law).

A reply. Suppose the law governing A-B interactions entails that m: in any A-B interaction, the two particles remain an A-on and a B-on afterward (at least until some time passes and one of them interacts with something else). Let ϕ now be: had an X-Y interaction occurred at time t, then *all* particles would have been X-ons or Y-ons after t (at least until any further interaction between them occurred). Now ϕ is logically consistent with the laws and conditionals in Λ°. For example, ϕ and m are logically consistent, though if both hold, then there cannot be an A-B interaction along with an X-Y interaction at t (since by m, an A-B interaction would have to leave an A-on and a B-on).

Therefore, "Had ϕ held, then m would still have held" must be true for Λ° (containing the laws and associated conditionals) to be Stable. It might be insisted that if m and its colleagues really are all of the laws, despite being incomplete, then this counterfactual is true—for had ϕ held, then either no X-Y interaction or no A-B interaction would have occurred at t.

But what about "Had ϕ held, then had an X-Y interaction occurred at t and an A-B interaction occurred at t, then m would still have held"? (Symbolically: $\phi \,\square\!\!\rightarrow (p \,\square\!\!\rightarrow m)$.) Since ϕ and p are separately logically consistent with m and its colleagues in Λ°, Λ°'s Stability requires that $\phi \,\square\!\!\rightarrow (p \,\square\!\!\rightarrow m)$ hold in all contexts. Let's grant that the 54 laws are so diverse that although there would have been an X-Y law had ϕ held, that law would have slotted in without disrupting any of the 54. So had ϕ, then m would still have been a law but there would also have been an X-Y interaction law demanding that all particles be X-ons or Y-ons in the wake of an X-Y interaction. So (as we saw) the resulting laws would have entailed that p is false—that it is not the case that an X-Y interaction and an A-B interaction both occur at t. So under ϕ, the counterfactual antecedent p is a counterlegal supposition. Had ϕ held, then had p held, either the "new" X-Y law would not still have held or m would not still have held. In this contest (of a kind we have seen before) between the "new" X-Y law and m, I see no reason why m must win in all contexts. So it is not the case that $\phi \,\Box\!\!\rightarrow (p \,\Box\!\!\rightarrow m)$ holds in all contexts, as required for $\Lambda°$'s Stability where the laws are incomplete.

(However, Carroll notes that not every incomplete set of laws is vulnerable to the maneuver I have just deployed. For instance, suppose that the only law is that everything has mass.)

79. The truth of "If it had been the case that the flip would have landed heads had I flipped the coin in L, then m would not still have held" is logically consistent with $\Lambda°$'s Stability since in those contexts where it holds, its antecedent ("flip $\Box\!\!\rightarrow$ heads") is logically inconsistent with "flip $\Box\!\!\rightarrow$ 50% chance heads." Hence, $\Lambda°$ (which includes "flip $\Box\!\!\rightarrow$ 50% chance heads," considering the law m) does not need to be preserved under the supposition that "flip $\Box\!\!\rightarrow$ heads" in order for $\Lambda°$ to be Stable. One way to show that in certain contexts, "flip $\Box\!\!\rightarrow$ heads" is logically inconsistent with "flip $\Box\!\!\rightarrow$ 50% chance heads," even though "heads" and "50% chance heads" are logically consistent, is to recall chapter 3: "flip $\Box\!\!\rightarrow$ 50% chance heads" logically entails "Had the coin been flipped, it might have landed tails," which in certain contexts contradicts "flip $\Box\!\!\rightarrow$ heads."

80. I raised this possibility in chapter 3, section 3.6, and discuss it further in my 2006.

81. Hutchison 1993: 320.

82. Goodman 1983: 31.

References

Airy, G. B. 1830. "On certain conditions under which a perpetual motion is possible," *Transactions of the Cambridge Philosophical Society* 3: 369–72.

Armstrong, David. 1978. *Universals and Scientific Realism*. Cambridge: Cambridge University Press.

Armstrong, David. 1983. *What Is a Law of Nature?* Cambridge: Cambridge University Press.

Armstrong, David. 1997. *A World of States of Affairs*. Cambridge: Cambridge University Press.

Arntzenius, Frank. 2000. "Are there really instantaneous velocities?" *The Monist* 83(2): 187–208.

Augustine. 1982. *The Literal Meaning of Genesis,* vol. 1, trans. John Hammond Taylor. New York: Newman Press.

Baumann, Gerd. 2000. *Symmetry Analysis of Differential Equations with Mathematica*. Berlin: Springer.

Beebee, Helen. 2000. "The non-governing conception of laws of nature," *Philosophy and Phenomenological Research* 61: 571–93.

Bennett, Jonathan. 1974. "Counterfactuals and possible worlds," *Canadian Journal of Philosophy* 4: 381–402.

Bennett, Jonathan. 1984. "Counterfactuals and temporal direction," *Philosophical Review* 93: 57–91.

Bennett, Jonathan. 2003. *A Philosophical Guide to Conditionals*. Oxford, UK: Clarendon.

Bergmann, Peter G. 1962. "The special theory of relativity," in *Encyclopedia of Physics,* vol. 4, ed. S. Flügge. Berlin: Springer-Verlag, pp. 109–202.

Bergmann, Peter G. 1980. Comment in "Open discussion following papers by S. Hawking and W. G. Unruh," in *Some Strangeness in the Proportion*, ed. Harry Woolf. Reading, MA: Addison-Wesley, pp. 156–58.

Berzi, Vittorio, and Vittorio Gorini. 1969. "Reciprocity principle and Lorentz transformations," *Journal of Mathematical Physics* 10: 1518–24.

Bigelow, John. 1988. *The Reality of Numbers*. Oxford, UK: Clarendon.

Bigelow, John, Brian Ellis, and Caroline Lierse. 1992. "The world as one of a kind: Natural necessity and laws of nature," *British Journal for the Philosophy of Science* 43: 371–88.

Bigelow, John, and Robert Pargetter. 1990. *Science and Necessity*. Cambridge: Cambridge University Press.

Blackburn, Simon. 1993. "Morals and modals," in Simon Blackburn, *Essays in Quasi-Realism*. New York: Oxford University Press, pp. 52–74.

Blaser, J. P. 2001. "Remarks by Heinrich Hertz (1857–94) on the equivalence principle," *Classical and Quantum Gravity* 18: 2393–95.

Born, Max. 1961. *Einstein's Theory of Relativity*. New York: Dover.

Braine, David. 1972. "Varieties of necessity," *Supplementary Proceedings of the Aristotelian Society* 46: 139–70.

Bulhof, Johannes. 1999. "What if? Modality and history," *History and Theory* 38: 145–68.

Carroll, John. 1994. *Laws of Nature*. Cambridge: Cambridge University Press.

Carroll, John. 2002. "Instantaneous motion," *Philosophical Studies* 110: 49–67.

Chisholm, Roderick. 1946. "The contrary-to-fact conditional," *Mind* 55: 289–307.

Chisholm, Roderick. 1955. "Law statements and counterfactual inference," *Analysis* 15: 97–105.

Comins, Neil. 1993. *What If the Moon Didn't Exist: Voyages to Earths That Might Have Been*. New York: HarperCollins.

Cruttwell, C. R. M. F. 1991. *A History of the Great War 1914–1918*. Chicago: Academy.

Darwin, Charles. 1868. *The Variation of Animals and Plants under Domestication,* vol. 2. London: John Murray.

Davies, Paul. 1995. "Algorithmic compressibility, fundamental and phenomenological laws," in *Laws of Nature: Essays on the Philosophical, Scientific, and Historical Dimensions,* ed. Friedel Weinert. Berlin: de Gruyter, pp. 248–67.

Descartes, René. 2000. *Philosophical Essays and Correspondence,* ed. Roger Ariew. Indianapolis, IN: Hackett.

Desmond, Adrian. 1997. *Huxley: From Devil's Disciple to Evolution's High Priest*. Reading, MA: Addison-Wesley.

Divers, John. 2002. *Possible Worlds*. London: Routledge.

Dorato, Mauro. 2005. *The Software of the Universe*. Aldershot, UK: Ashgate.

Downing, P. B. 1959. "Subjunctive conditionals, time order, and causation," *Proceedings of the Aristotelian Society* 59: 125–40.

Dretske, Fred. 1977. "Laws of nature," *Philosophy of Science* 44: 248–68.

Earman, John. 1986. *A Primer on Determinism*. Dordrecht, Netherlands: Reidel.

Earman, John. 1989. *World Enough and Space-Time*. Cambridge, MA: MIT Press.

Earman, John. 1995. *Bangs, Crunches, Whimpers, and Shrieks*. New York: Oxford University Press.

Eddington, Arthur. 1928. *The Nature of the Physical World*. New York: Macmillan.

Ehrenfest, Paul. 1917. "In what way does it become manifest in the fundamental laws of physics that space has three dimensions?" *Proceedings of the Amsterdam Academy* 20: 200–9.

Einstein, Albert. 1935. "Elementary derivation of the equivalence of mass and energy," *Bulletin of the American Mathematical Society* 41: 223–30.

Einstein, Albert. 1954. *Ideas and Opinions*. New York: Bonanza.

Einstein, Albert. 1961. *Relativity: The Special and the General Theory*. New York: Crown.

Einstein, Albert, and Leopold Infeld. 1951. *The Evolution of Physics*. New York: Simon and Schuster.

Ellis, Brian. 1999. "Causal powers and laws of nature," in *Causation and Laws of Nature*, ed. Howard Sankey. Dordrecht, Netherlands: Kluwer, pp. 19–34.

Ellis, Brian. 2001. *Scientific Essentialism*. Cambridge: Cambridge University Press.

Ellis, Brian. 2002. *The Philosophy of Nature: A Guide to the New Essentialism*. Montreal and Kingston: McGill-Queen's University Press.

Ellis, Brian. 2005. "Marc Lange on essentialism," *Australasian Journal of Philosophy* 83: 75–79.

Emerson, William. 1768. *The Doctrine of Fluxions,* 3rd ed. London: Robinson and Roberts.

Faraday, Michael. 1858. "On Wheatstone's electric telegraph in relation to science," in *Notices of the Proceedings at the Meetings of the Members of the Royal Institution of Great Britain,* vol. 2. London: William Clowes, pp. 555–60.

Feinberg, Gerald, and Maurice Goldhaber. 1963. "The conservation laws of physics," *Scientific American* 209 (October): 36–45.

Feynman, Richard. 1967. *The Character of Physical Law*. Cambridge, MA: MIT Press.

Feynman, Richard, Robert Leighton, and Matthew Sands. 1963. *The Feynman Lectures on Physics*. Reading, MA: Addison-Wesley.

Field, Hartry. 1989. *Realism, Mathematics, and Modality*. Oxford, UK: Blackwell.

Foster, John. 2004. *The Divine Lawmaker*. Oxford, UK: Clarendon.

Gardner, Martin. 1964. *The Ambidextrous Universe*. New York: Basic Books.

Gardner, Martin. 1970. "Mathematical games: The fantastic combinations of Conway's new solitaire game 'life,'" *Scientific American* 223 (October): 120–23.

Goodman, Nelson. 1947. "The problem of counterfactual conditionals," *The Journal of Philosophy* 44: 13–28.

Goodman, Nelson. 1983. *Fact, Fiction, and Forecast,* 4th ed. Cambridge, MA: Harvard University Press.

Gross, David. 1996. "The role of symmetry in fundamental physics," *Proceedings of the National Academy of Sciences USA* 93: 14256–59.

Haavelmo, Trygve. 1944. "The probability approach to econometrics," *Econometrica* 12 (suppl.): 1–117.

Hamilton, William Rowan. 1834. "On a general method in dynamics," *Philosophical Transactions of the Royal Society,* part II, 247–308, repr. *The Mathematical Papers of Sir William Rowan Hamilton*. Cambridge: Cambridge University Press, 1940, pp. 103–61.

Handfield, Toby. 2005. "Lange on essentialism, counterfactuals, and explanation," *Australasian Journal of Philosophy* 83: 81–85.

Hempel, C. G. 1965. *Aspects of Scientific Explanation and Other Essays in the Philosophy of Science*. New York: Free Press.

Hempel, C. G. 1966. *Philosophy of Natural Science*. Englewood Cliffs, NJ: Prentice-Hall.

Hempel, C. G., and Paul Oppenheim. 1948. "Studies in the Logic of Explanation," *Philosophy of Science* 15: 135–75.

Hooker, Richard. 1632. *Of the Lawes of Ecclesiastical Politie*. London: William Standsbye.

Horwich, Paul. 1987. *Asymmetries in Time*. Cambridge, MA: MIT Press.

Houtappel, R. M. F., H. Van Dam, and E. P. Wigner. 1965. "The conceptual basis and use of the geometric invariance principles," *Reviews of Modern Physics* 37: 595–632.

Hunt, Bruce. 1991. *The Maxwellians*. Ithaca, NY: Cornell University Press.

Hutchison, Keith. 1993. "Is classical physics really time-reversible and deterministic?" *British Journal for the Philosophy of Science* 44: 307–23.

Hutton, Charles. 1796. *Mathematical and Philosophical Dictionary*. London: J. Davis for J. Johnson.

Huxley, Thomas Henry. 1871. "A liberal education," in *Lay Sermons, Addresses, and Reviews*. New York: Appleton.

Jackson, Frank. 1977. "A causal theory of counterfactuals," *Australasian Journal of Philosophy* 55: 3–21.

Jackson, Frank, and Robert Pargetter. 1988. "A question about rest and motion," *Philosophical Studies* 53: 141–46.

Jesseph, Douglas. 1993. *Berkeley's Philosophy of Mathematics*. Chicago: University of Chicago Press.

Kim, Seahwa, and Cei Maslen. 2006. "Counterfactuals as short stories," *Philosophical Studies* 129: 81–117.

Knott, Cargill Gilston. 1911. *Life and Scientific Work of Peter Guthrie Tait*. Cambridge: Cambridge University Press.

Kratzer, Angelika. 1991. "Modality," in *Semantics: An International Handbook of Contemporary Research,* ed. A. von Stechow and D. Wunderlich. Berlin: de Gruyter, pp. 639–50.

Lagrange, Joseph Louis. 1811. *Analytical Mechanics,* repr. Boston Studies in the Philosophy of Science, vol. 191, trans. A. Boissonnade and V. Vagliente. Dordrecht, Netherlands: Kluwer, 1997.

Landau, L. D., and E. M. Lifshitz. 1976. *Mechanics,* 3rd ed. Oxford, UK: Pergamon.

Lange, Marc. 2000. *Natural Laws in Scientific Practice*. New York: Oxford University Press.

Lange, Marc. 2002a. *An Introduction to the Philosophy of Physics: Locality, Fields, Energy, and Mass*. Malden, MA: Blackwell.

Lange, Marc. 2002b. "Who's afraid of ceteris-paribus laws? Or: How I learned to stop worrying and love them," *Erkenntnis* 57: 407–23.

Lange, Marc. 2004. "The autonomy of functional biology: A reply to Rosenberg," *Biology and Philosophy* 19: 93–101.

Lange, Marc. 2005. "A counterfactual analysis of the concepts of logical truth and necessity," *Philosophical Studies* 125: 277–303.

Lange, Marc. 2006. "Do chances receive equal treatment under the laws? Or: Must chances be probabilities?" *British Journal for the Philosophy of Science* 57: 383–403.

Laraudogoitia, Jon Perez. 1996. "A beautiful supertask," *Mind* 105: 81–83.

Leeds, Stephen. 2001. "Possibility: Physical and metaphysical," *Physicalism and Its Discontents*, ed. Carl Gillett and Barry Loewer. Cambridge: Cambridge University Press, pp. 172–93.

Leibniz, G. W. 1902. *Discourse on Metaphysics*. LaSalle, IL: Open Court.

Lévy-Leblond, Jean-Marc. 1976. "One more derivation of the Lorentz transformations," *American Journal of Physics* 44: 271–77.

Lewis, David. 1973. *Counterfactuals*. Cambridge, MA: Harvard University Press.

Lewis, David. 1983. "New work for a theory of universals," *Australasian Journal of Philosophy* 61: 343–77.

Lewis, David. 1986a. *On the Plurality of Worlds*. Oxford, UK: Blackwell.

Lewis, David. 1986b. *Philosophical Papers: Volume II*. New York: Oxford University Press.

Lewis, David. 1999a. "Humean supervenience debugged," in David Lewis, *Papers in Metaphysics and Epistemology*. Cambridge: Cambridge University Press, pp. 224–47.

Lewis, David. 1999b. "A world of truthmakers," in David Lewis, *Papers in Metaphysics and Epistemology*. Cambridge: Cambridge University Press, pp. 215–21.

Linde, Andrei. 1994. "The self-reproducing, inflationary universe," *Scientific American* 271 (November): 48–55.

Loewer, Barry. 2004. "David Lewis's Humean theory of objective chance," *Philosophy of Science* 71: 1115–25.

Lycan, W. G. 1994. "Relative modality," in W. G. Lycan, *Modality and Meaning*. Dordrecht, Netherlands: Kluwer, pp. 171–200.

Mackie, J. L. 1962. "Counterfactuals and causal laws," in *Analytic Philosophy*, ed. R. S. Butler. New York: Barnes and Noble, pp. 66–80.

Maclaurin, Colin. 1742. *A Treatise of Fluxions*. Edinburgh, UK: Ruddimans.

Maclaurin, Colin. 1748/1971. *An Account of Sir Isaac Newton's Philosophical Discoveries*. Hildesheim, Germany: Olms Verlag.

Maudlin, Tim. 2007. *The Metaphysics within Physics*. Oxford: Oxford University Press.

Meyer, Ulrich. 2003. "The metaphysics of velocity," *Philosophical Studies* 112: 93–102.

Mill, John Stuart. 1874. *A System of Logic*, 8th ed. New York: Harper and Bros.

Nahin, Paul. 1987. *Oliver Heaviside: Sage in Solitude*. New York: IEEE Press.

Nambu, Yoichiro. 1985. "Directions in particle physics," *Progress in Theoretical Physics* (Supplement) 85: 104–10.

Nordenskiold, Erik. 1936. *The History of Biology*. New York: Tudor.

Norton, John. 2007. "Causation as folk science," in *Causation and the Constitution of Reality*, ed. Huw Price and Richard Corry. Oxford: Oxford University Press, pp. 11–44.

Pais, Abraham. 1986. *Inward Bound*. Oxford, UK: Clarendon.

Pal, Palash. 2003. "Nothing but Relativity," *European Journal of Physics* 24: 315–19.

Papineau, David. 2001. "The rise of physicalism," in *Physicalism and Its Discontents*, ed. Carl Gillett and Barry Loewer. Cambridge: Cambridge University Press, pp. 3–36.

Penrose, Roger. 1987. "Newton, quantum theory, and reality," in *300 Years of Gravitation*, ed. Stephen Hawking and Werner Israel. Cambridge: Cambridge University Press, pp. 17–49.

Poincaré, Henri. 1913/1963. "The evolution of laws," in *Mathematics and Science: Last Essays*, trans. John W. Bolduc. New York: Dover, pp. 1–14.

Pollock, John. 1974. "Subjunctive generalizations," *Synthese* 28: 199–214.

Pollock, John. 1976. *Subjunctive Reasoning*. Dordrecht, Netherlands: Reidel.

Putnam, Hilary. 1990. *Realism with a Human Face*. Cambridge, MA: Harvard University Press.

Quine, W.V. O. 1960. *Word and Object*. New York: Wiley.

Quine, W.V. O. 1961. *From a Logical Point of View*, 2nd rev. ed. Cambridge, MA: Harvard University Press.

Quine, W.V. O. 1970. *Philosophy of Logic*. Englewood Cliffs, NJ: Prentice Hall.

Ransom, Roger. 2005. *The Confederate States of America: What Might Have Been*. New York: Norton.

Reichenbach, Hans. 1947. *Elements of Symbolic Logic*. New York: Macmillan.

Reichenbach, Hans. 1954. *Nomological Statements and Admissible Operations*. Dordrecht, Netherlands: North-Holland.

Rocci, Andrea. 2005. "On the nature of the epistemic readings of the Italian modal verbs: The relationship between propositionality and inferential discourse relations," in *Crosslinguistic Views on Tense, Aspect and Modality*, ed. A. van Hout, B. Hollebrandse, and C.Vet. Amsterdam: Rodopi, pp. 229–46.

Russell, Bertrand. 1903/1937. *Principles of Mathematics*, 2nd ed. London: G. Allen & Unwin.

Russell, Bertrand. 1917. *Mysticism and Logic*. London: George Allen & Unwin.

Russell, Bertrand. 1919. *Introduction to Mathematical Philosophy*. New York and London: George Allen & Unwin.

Salmon, Wesley. 1980. *Space, Time, and Motion*, 2nd ed. Minneapolis: University of Minnesota Press.

Salmon, Wesley. 1984. *Scientific Explanation and the Causal Structure of the World*. Princeton, NJ: Princeton University Press.

Schaffer, Jonathan. 2003. "Principled chances," *British Journal for the Philosophy of Science* 54: 27–41.

Searle, G. F. C. 1897. "On the steady motion of an electrified ellipsoid," *Philosophical Magazine* (ser. 5) 44: 329–41.

Seelau, Eric P., Sheila M. Seelau, Gary L. Wells, and Paul D. Windschitl. 1995. "Counterfactual constraints," in *What Might Have Been: The Social Psychology of Counterfactual Thinking*, ed. Neil J. Roese and James M. Olson. Hillsdale, NJ: Lawrence Erlbaum, pp. 57–80.

Shoemaker, Sydney. 1998. "Causal and metaphysical necessity," *Pacific Philosophical Quarterly* 79: 59–77.

Spinoza, Benedict. 1951. *Theological-Political Treatise*, trans. R. H. M. Elwes. New York: Dover.

Stalnaker, Robert. 1968. "A theory of conditionals," *American Philosophical Quarterly Monographs* 2 (*Studies in Logical Theory*): 98–112.

Stalnaker, Robert. 1984. *Inquiry*. Cambridge, MA: MIT Press.

Stöltzner, Michael. 1995. "Levels of physical theories," in *The Foundational Debate*, Vienna Circle Institute Yearbook 3, ed. Werner DePauli-Schimanovich, Eckehart Köhler, and Friedrich Stadler. Dordrecht, Netherlands: Kluwer, pp. 47–64.

Strawson, P. F. 1952. *Introduction to Logical Theory*. London: Methuen.

Swartz, Norman. 1985. *The Concept of Physical Law*. Cambridge: Cambridge University Press.

Tarski, Alfred. 1956. *Logic, Semantics, Metamathematics*, trans. J. H. Woodger. Oxford, UK: Clarendon.

Thirring, Walter. 1995. "Do the laws of nature evolve?" in *What Is Life? The Next Fifty Years*, ed. Michael P. Murphy and Luke A. J. O'Neill. Cambridge: Cambridge University Press, pp. 131–36.

Thomson, Judith Jarvis. 1990. *The Realm of Rights*. Cambridge, MA: Harvard University Press.

Thomson, William, and Peter Guthrie Tait. 1888. *Treatise on Natural Philosophy*. Cambridge: Cambridge University Press.

Todd, William. 1964. "Counterfactual conditionals and the presuppositions of induction," *Philosophy of Science* 31: 101–10.

Tooley, Michael. 1977. "The nature of laws," *Canadian Journal of Philosophy* 7: 667–98.

Tooley, Michael. 1987. *Causation: A Realist Approach*. Oxford, UK: Clarendon.

Tooley, Michael. 1988. "In defense of the existence of states of motion," *Philosophical Topics* 16: 225–54.

Upgren, Arthur. 2005. *Many Skies: Alternative Histories of the Sun, Moon, Planets, and Stars*. New Brunswick, NJ: Rutgers University Press.

Van Fraassen, Bas. 1977. "The pragmatics of explanation," *American Philosophical Quarterly* 4: 143–50.

Van Fraassen, Bas. 1980. *The Scientific Image*. Oxford, UK: Clarendon.

Van Fraassen, Bas. 1984. "Belief and the will," *The Journal of Philosophy* 81: 235–56.

Van Fraassen, Bas. 1989. *Laws and Symmetry*. Oxford, UK: Clarendon.

Van Inwagen, Peter. 1979. "Laws and counterfactuals," *Nous* 13: 439–53.

Von Fintel, Kai. 2006. "Modality and language," in *Encyclopedia of Philosophy*, 2nd ed., ed. Donald Borchert. Detroit: Macmillan Reference USA, vol. 10, pp. 20–27.

Walton, John. 1735. *The Catechism of the Author of the Minute Philosopher Fully Answer'd*, 2nd ed. Dublin: S. Powell for William Smith.

Weinberg, Steven. 1977. *The First Three Minutes*. New York: Basic Books.

Weinberg, Steven. 1992. *Dreams of a Final Theory*. New York: Pantheon.

Wertheimer, Roger. 1972. *The Significance of Sense*. Ithaca, NY: Cornell University Press.

Wigner, Eugene 1954a. "Conservation laws in classical and quantum physics," *Progress of Theoretical Physics* 11: 437–40.

Wigner, Eugene. 1954b. "On kinematic and dynamic laws of symmetry," in *Symposium on New Research Techniques in Physics, July 15–29, 1952*. Rio de Janeiro: Academia Brasileira de Ciências, pp. 199–200.

Wigner, Eugene. 1964. "Symmetry and conservation laws," *Proceedings of the National Academy of Sciences USA* 51: 956–65.

Wigner, Eugene. 1972. "Events, laws of nature, and invariance principles," in *Nobel Lectures: Physics 1963–1970*. Amsterdam: Elsevier, pp. 6–19.

Wigner, Eugene. 1984. "Changes in the roles of the symmetry principles," in *13th International Colloquium on Group Theoretical Methods in Physics*, ed. W. W. Zachary. Singapore: World Scientific, pp. 591–97.

Wigner, Eugene. 1985. "Events, laws of nature, and invariance principles," in *How Far Are We from the Gauge Forces—Proceedings of the 21st Course of the International School of Subnuclear Physics, Aug 3–14, 1983, Enrice Sicily*, ed. A. Zichichi. New York and London: Plenum, pp. 699–708.

Wigner, Eugene. 1997. "Events, laws of nature, and invariance principles," in *Collected Papers*, part A, vol. 3, ed. Arthur Wightman. Berlin: Springer, pp. 185–96.

Wilczek, Frank. 2004. "Whence the force of F = ma? I. Culture shock," *Physics Today* 57 (October): 11–12.

Williamson, Timothy. 2005. "Armchair philosophy, metaphysical modality and counterfactual thinking," *Proceedings of the Aristotelian Society* 105: 1–23.

Yang, C. N. 1964. "The law of parity conservation and other symmetry laws of physics," in *Nobel Lectures: Physics 1942–1962,* ed. Bengt Samuelson and Michael Sohlman. Amsterdam: Elsevier, pp. 393–403.

Zemanian, A. H. 1965. *Distribution Theory and Transform Analysis.* New York: McGraw-Hill.

Index